John A. Roebling's Sons Company

Wire in Electrical Construction

John A. Roebling's Sons Company

Wire in Electrical Construction

ISBN/EAN: 9783337214562

Printed in Europe, USA, Canada, Australia, Japan

Cover: Foto ©berggeist007 / pixelio.de

More available books at **www.hansebooks.com**

WIRE

IN

Electrical Construction

John A. Roebling's Sons Co.

TRENTON, N. J.

117-119 Liberty street,　　171-173 Lake street,
　NEW YORK.　　　　　　　　CHICAGO.

32 South Water street,　　25-27 Fremont street,
　CLEVELAND.　　　　　　　SAN FRANCISCO.

The Brandt Press,
Trenton.
1897.

COPYRIGHTED, 1897,
BY
JOHN A. ROEBLING'S SONS CO.

All rights reserved.

PREFATORY.

THE OBJECT of this book is to give in a convenient form the properties and dimensions of bare and insulated wires and cables used in electrical construction. No attempt has been made to describe the uses of wire in any of the applications of electricity. To go into this would require that the whole field of electrical engineering be covered.

It is believed that some of the matter is new. All of the tables have been very carefully computed, and are believed to be correct.

In nearly all cases the formulæ and constants used in computing tables are given, so that the user can determine at once the basis from which the table was calculated. A considerable amount of work has been done in testing samples to determine the proper constants. In many cases this has taken more time than the actual preparation of the tables.

It is hoped that the work will be acceptable to the users of electrical wires, and that some of the labor involved in the preparation of these tables will be saved to those using the book.

JOHN A. ROEBLING'S SONS CO.

TRENTON, N. J., May, 1897.

TABLE OF CONTENTS.

	PAGE.
MEASURES AND THEIR EQUIVALENTS:	
Measures of length..	1
Measures of area..	2
Measures of volume...	3
Measures of weight..	4
Measures of work...	5
Measures of pressure...	6
Decimal equivalents of parts of an inch.......................	7
Wire gauges in mils...	8
Wire gauges in millimeters..	9
Tables of specific gravities:	
Metals..	10
Liquids..	11
Gases...	11
Weights of substances..	12
The comparison of thermometers:	
Fahrenheit to Centigrade.......................................	13
Centigrade to Fahrenheit.......................................	13
Electrical units...	14–15
COPPER WIRE:	
Formulæ and explanations...	16–17
Matthiessen's standard...	17
Temperature coëfficients..	18
Properties of copper wire—weights, resistances, etc.	
English system:	
Brown & Sharpe gauge..	19
Birmingham wire gauge.......................................	20
New British standard gauge................................	21
Metric system:	
Brown & Sharpe gauge..	22
Weights of all gauges...	23
Hard-drawn copper wire:	
British Post-office specifications............................	24
Telephone specifications...	25
Tensile strength of copper wire...................................	26
Bi-metallic wire...	27
Strands of copper wire:	
Formulæ and explanations......................................	28
Diameters and properties..	29
Diameters of wires in strands.................................	30–31
Numbers of wires in strands...................................	32–33

TABLE OF CONTENTS.

IRON WIRE: PAGE.
- Formulæ and explanations.. 34
- Properties of iron wire—weights, strength, resistances, etc., 35
- Specifications:
  - Western Union Telegraph company........................ 36
  - British Post-office... 37
- Strands:
  - Formulæ and explanations....................................... 38
  - Properties of galvanized steel wire strands—weights and breaking strength................................. 38
- Supporting capacity of galvanized strands...................... 39

CURRENTS:
- Fusing effects:
  - Diameters of wires... 40
  - Current required.. 41
- Heating effects:
  - References and explanations................................. 42
  - Carrying capacity:
    - Insurance rules... 43
    - Insulated wires in mouldings....................... 44
    - Wires indoors... 45
    - Wires outdoors.. 46

SPANS:
- Formulæ and explanations.. 47–49
- Specifications... 48
- Strains at centers of spans... 50–52
- Total lengths of wires in spans..................................... 54–55
- Deflections in spans at various temperatures............... 53

DESCRIPTION OF THE ROEBLING ELECTRIC WIRES:
- Weatherproof wires... 56–57
- Rubber wires.. 58–59
- Magnet wire... 60–61
- German silver wire... 62
- Office wires.. 63
- Cables:
  - Power cables... 64–65
  - Telephone cables... 66–67
  - Telegraph cables... 68–69
  - Aerial cables... 70–71
  - Submarine cables.. 72
- Rail-bonds.. 73

MEASURES OF LENGTH.

Names of units.	Inches.	Feet.	Yards.	Meters.	Chains.	Kilometers.	Miles.	Knots.
Inches............	1.	.083 33	.027 78	.025 4	.001 26	.000 025	.000 015 8	.000 013 7
Feet..............	12.	1.	.333 33	*.304 801	.015 15	.000 305	.000 189	.000 164 5
Yards.............	36.	3.	1.	*.914 402	.045 45	.000 914	.000 568	.000 493 4
Meters............	*39.37	*3.280 83	*1.093 611	1.	.049 71	.001	.000 621	.000 54
Chains............	792.	66.	22.	20.116 9	1.	.020 116 9	.012 5	.010 855
Kilometers........	39 370.	3 280.83	1 093.61	1 000.	49.71	1.	*.621 87	.539 61
Miles.............	63 360.	5 280.	1 760.	1 609.85	80.	*1.609 35	1.	.868 42
Knots.............	72 960.	6 080.	2 026.66	1 853.19	92.112	1.853 19	1.151 5	1.

Mil = .001 inch.

In these tables the equivalents of the metric system of weights and measures are those given in 1890 by the United States Coast and Geodetic Survey, Office of Standard Weights and Measures. These values in all the tables are marked by an asterisk (*). The other equivalents are calculated from these.

In the metric system the following prefixes are used for subdivisions and multiples:

Mill = $\frac{1}{1000}$ Deca = 10
Centi = $\frac{1}{100}$ Hecto = 100
Deci = $\frac{1}{10}$ Kilo = 1 000
                           Myria = 10 000

MEASURES OF AREA.

Names of units.	Circular mils.	Square mils.	Square millimeters.	Square centimeters.	Square inches.	Square feet.	Square yards.	Square meters.
Circular mils......	1.	.785 4	.000 506 7	....	....	....	....	....
Square mils........	1.273 2	1.	.000 645	.000 006 4	.000 001	....	....	....
Square millimeters.	1 973.5	1 550.1	1.	.01	.001 55	....	....	....
Square centimeters.	197 350.	155 010.	100.	1.	*.155	.001 077	.000 12	.000 1
Square inches.......	1 273 239.	1 000 000.	645.2	*6.452	1.	.006 94	.000 77	.000 645
Square feet.........	....	....	92 900.	929.	144.	1.	.111 11	.092 9
Square yards........	....	....	836 100.	8 361.	1 296.	9.	1.	*.836
Square meters.......	....	....	1 000 000.	10 000.	1 550.016	*10.764	*1.196	1.

Circular mil = a circle whose diameter is .001 inch.
Square mil = a square whose sides are .001 inch.

MEASURES OF VOLUME.

Names of units.	Cubic centimeters.	Cubic inches.	Liters.	Gallons.	Cubic feet.	Cubic yards.	Cubic meters.
Cubic centimeters......	1.	*.061	.001	.000 264	.000 035	.000 001 3	.000 001
Cubic inches	*16.387	1.	.016 887	.004 33	.000 578	.000 021 4	.000 016
Liters..................	1 000.	61.023	1.	*.264 17	.035 314	.001 308	.001
Gallons................	3 785.4	231.	*3.785 44	1.	.133 68	.004 952	.003 785
Cubic feet.............	28 315.	1 728.	28.315	7.48	1.	.037 037	*.028 32
Cubic yards...........	764 552.	46 656.	764.55	201.97	27.	1.	*.765
Cubic meters	1 000 000.	61 023.	1 000.	264.17	*35.314	*1.308	1.

Fluid ounce = 29.57 cubic centimeters.
Gallon = 128 fluid ounces.
Gallon = 4 quarts.
Quart = 2 pints.

MEASURES OF WEIGHT.

Names of units.	Grains.	Grams.	Ounces avoirdupois.	Pounds troy.	Pounds avoirdupois.	Kilograms.
Grains	1.	.064 798 9	.002 28	.000 174	.000 143	.000 064
Grams	*15.432	1.	.035 27	.002 68	.002 205	.001
Ounces avoirdupois	437.5	*28.349 5	1.	.075 95	.062 5	.028 35
Pounds troy	5 760.	373.24	13.166	1.	.822 86	.373 24
Pounds avoirdupois	7 000.	453.59	16.	1.215 3	1.	*.453 59
Kilograms	*15 432.36	1 000.	35.274	2.679 2	*2.204 62	1.
Long tons	15 680 000.	1 016 041.6	35 840.	2 722.2	2 240.	1 016.04

1 kilogram per kilometer = .671 95 pounds per 1 000 feet.
1 pound per thousand feet = 1.488 2 kilograms per kilometer.

MEASURES OF WORK.

Names of units.	Ergs.	Gram-degree Centigrade.	Pound-degree Fahrenheit.	Watt-second.	Kilogram-meter.	Foot-pound.	Horse-power-second.
Gram-degree Centigrade......	41 549 500.	1.	.003 968 3	4.154 95	.423 54	3.063 5	.005 57
Pound-degree Fahrenheit.....	10 470 300 000.	252.11	1.	1 047.08	106.781	772.	1.403
Watt-second	10 000 000.	.240 7	.000 955 1	1.	.101 987	.737 324	.001 340 6
Kilogram-meter.	98 100 000.	2.361	.009 369	9.81	1.	7.233 14	.013 151
Foot-pound......	13 562 600.	.326 4	.001 295 3	1.356 26	.138 25	1.	.001 818 18
Horse-power-second	....	179.5	.712 4	745.94	76.089	550.	1.

Joule = volt-coulomb = watt for one second.
Calorie = gram-degree Centigrade.
B. T. U. = British thermal unit = pound-degree Fahrenheit.

MEASURES OF PRESSURE.

Names of units.	Atmospheres.	Pounds on square inch.	Inches of mercury at 32° F.	Feet of water at 60° F.	Millimeters of mercury at 32° F.	Pounds on square foot.	Kilograms on square meter.
Atmospheres............................	1.	14.7	29.922	33.94	760.	2 116.	10 333.
Pounds on square inch............	.068 08	1.	2.036	2.309	51.7	143.946	702.925
Inches of mercury at 32° F......	.033 42	.491 3	1.	1.134	25.398	70.7	345.331
Feet of water at 60° F.............	.029 47	.433 2	.881 8	1.	22.399	62.35	304.565
Millimeters of mercury at 32° F....	.001 316	.019 34	.089 37	.044 64	1	2.784	13.596
Pounds on square foot............	.000 472 6	.006 947	.014 14	.016 03	.359 2	1.	4.883
Kilograms on square meter.....	.000 096 77	.001 423	.002 896	.003 283	.073 65	.204 8	1.

1 kilogram per square millimeter = 1423 pounds per square inch.
1 pound per square inch = .000 703 kilograms per square millimeter.

DECIMAL EQUIVALENTS OF PARTS OF AN INCH.

16ths.	32ds.	64ths.	Mils.	16ths.	32ds.	64ths.	Mils.
		1	15.625			33	515.625
	1	2	31.25		17	34	531.25
		3	46.875			35	546.875
1	2	4	62.5	9	18	36	562.5
		5	78.125			37	578.125
	3	6	93.75		19	38	593.75
		7	109.875			39	609.375
2	4	8	125.	10	20	40	625.
		9	140.625			41	640.625
	5	10	156.25		21	42	656.25
		11	171.875			43	671.875
3	6	12	187.5	11	22	44	687.5
		13	203.125			45	703.125
	7	14	218.75		23	46	718.75
		15	234.375			47	734.375
4	8	16	250.	12	24	48	750.
		17	265.625			49	765.625
	9	18	281.25		25	50	781.25
		19	296.875			51	796.875
5	10	20	312.5	13	26	52	812.5
		21	328.125			53	828.125
	11	22	343.75		27	54	843.75
		23	359.375			55	859.375
6	12	24	375.	14	28	56	875.
		25	390.625			57	890.625
	13	26	406.25		29	58	906.25
		27	421.875			59	921.875
7	14	28	437.5	15	30	60	937.5
		29	453.125			61	953.125
	15	30	468.75		31	62	968.75
		31	484.375			63	984.375
8	16	32	500.	16	32	64	1 000.

WIRE GAUGES IN MILS.

Numbers.	Roebling.	Brown & Sharpe.	Birmingham or Stubs.	New British standard.
000 000	460.	....	....	464.
00 000	430.	....	....	432.
0 000	393.	460.	454.	400.
000	362.	409.6	425.	372.
00	331.	364.8	380.	348.
0	307.	324.9	340.	324.
1	283.	289.3	300.	300.
2	263.	257.6	284.	276.
3	244.	229.4	259.	252.
4	225.	204.3	238.	232.
5	207.	181.9	220.	212.
6	192.	162.	203.	192.
7	177.	144.3	180.	176.
8	162.	128.5	165.	160.
9	148.	114.4	148.	144.
10	135.	101.9	134.	128.
11	120.	90.74	120.	116.
12	105.	80.81	109.	104.
13	92.	71.96	95.	92.
14	80.	64.08	83.	80.
15	72.	57.07	72.	72.
16	63.	50.82	65.	64.
17	54.	45.26	58.	56.
18	47.	40.3	49.	48.
19	41.	35.89	42.	40.
20	35.	31.96	35.	36.
21	32.	28.46	32.	32.
22	28.	25.35	28.	28.
23	25.	22.57	25.	24.
24	23.	20.1	22.	22.
25	20.	17.9	20.	20.
26	18.	15.94	18.	18.
27	17.	14.2	16.	16.4
28	16.	12.64	14.	14.8
29	15.	11.26	13.	13.6
30	14.	10.03	12.	12.4
31	13.5	8.93	10.	11.6
32	13.	7.95	9.	10.8
33	11.	7.08	8.	10.
34	10.	6.3	7.	9.2
35	9.5	5.62	5.	8.4
36	9.	5.	4.	7.6

WIRE GAUGES IN MILLIMETERS.

Numbers.	Roebling.	Brown & Sharpe.	Birmingham or Stubs.	New British standard.
000 000	11.683	....	....	11.785
00 000	10.921	....	....	10.972
0 000	9.982	11.683	11.531	10.16
000	9.195	10.404	10.794	9.448
00	8.407	9.266	9.652	8.839
0	7.798	8.251	8.636	8.229
1	7.188	7.348	7.62	7.62
2	6.68	6.544	7.213	7.01
3	6.198	5.827	6.579	6.401
4	5.715	5.19	6.045	5.893
5	5.257	4.621	5.588	5.385
6	4.877	4.115	5.156	4.877
7	4.496	3.665	4.572	4.47
8	4.115	3.263	4.191	4.064
9	3.759	2.906	3.759	3.657
10	3.429	2.588	3.404	3.251
11	3.048	2.305	3.048	2.947
12	2.667	2.052	2.768	2.641
13	2.337	1.828	2.413	2.337
14	2.032	1.628	2.108	2.032
15	1.829	1.449	1.829	1.829
16	1.6	1.291	1.651	1.626
17	1.372	1.15	1.473	1.422
18	1.194	1.024	1.245	1.219
19	1.041	.911 6	1.067	1.016
20	.889	.811 8	.889	.914 4
21	.812 8	.722 9	.812 8	.812 8
22	.711 2	.643 8	.711 2	.711 2
23	.635	.573 3	.635	.609 6
24	.584 2	.510 5	.558 8	.558 8
25	.508	.454 6	.508	.508
26	.457 2	.404 9	.457 2	.457 2
27	.431 8	.360 5	.406 4	.416 6
28	.406 4	.321 1	.355 6	.375 9
29	.381	.285 9	.330 2	.345 4
30	.355 6	.254 5	.304 8	.315
31	.342 9	.226 7	.254	.294 6
32	.330 2	.201 9	.228 6	.274 8
33	.279 4	.179 8	.203 2	.254
34	.254	.160 1	.177 8	.233 7
35	.241 3	.142 6	.127	.213 4
36	.228 6	.127	.101 6	.193

TABLES OF SPECIFIC GRAVITIES.

Metals.

Names of metals.	Specific gravity.	Weights per cubic foot.	Specific heat.	Melting point in degrees Fahrenheit.
Aluminum, cast.............	2.5[1]	156.06	.214 3	....
" hammered.	2.67[1]	166.67	....	....
Antimony...................	6.702[2]	418.37	.050 8	810.
Arsenic......................	5.763[3]	359.76	.081 4	365.
Barium.......................	4.[3]	249.7	....	....
Bismuth.....................	9.822[2]	613.14	.030 8	497.
Cadmium	8.604[4]	537.1	.056 7	500.
Calcium	1.566[4]	97.76	....	....
Chromium...................	7.3[6]	455.7	....	....
Cobalt	8.6	536.86	.107	....
Copper......................	8.895[7]	555.27	.095 1	1 996.
" rolled	8.878[2]	554.21	....	....
" cast...................	8.788[2]	548.59	....	....
" drawn................	8.946 3[2]	558.47	....	....
" hammered.........	8.958 7[5]	559.25	....	....
" pressed.............	8.931[9]	557.52	....	....
" electrolytic.........	8.914[6]	556.46	....	....
Gold.........................	19.258[2]	1 202.18	.032 4	2 016.
Iron, bar....................	7.483[9]	467.18	.13	2 786.
" wrought...............	.7.79	486.29	.113	3 286.
Steel........................	7.85	490.03	.116	3 286.
Lead........................	11 445[10]	714.45	.031 4	612.
Magnesium................	2.24[11]	139.83	.249 9	....
Manganese...............	6.9[12]	430.73	.114	3 000.
Mercury....................	13.568[13]	846.98	.031 9	— 38.
Nickel......................	7.832	488.91	.109 1	280 0.
Platinum...................	20.3[2]	1 267.22	.032 4	328 6.
Potassium.................	.865[14]	54.	.169 6	136.
Silver.......................	10.522[11]	656.84	.057	1 873.
Sodium....................	.972[14]	60.68	.293 4	194.
Strontium..................	2.504[4]	156.31	....	....
Tin..........................	7.291[2]	455.14	.056 2	442.
Zinc........................	6.861[2]	428.29	.095 5	773.

1. Wöhler.
2. Brisson.
3. Clarke.
4. Matthiessen.
5. Stromeyer.
6. Bunsen.
7. Hatchett.
8. Brezenius.
9. Marchand & Scheerer.
10. Musschenbroek.
11. Playfair & Joule.
12. Bergman.
13. Watts' Dictionary.
14. Gay-Lussac & Thenard.

TABLES OF SPECIFIC GRAVITIES.—(Cont.)
Liquids.

Names of liquids.	Specific gravity.	Temperatures.
Alcohol	0.815 71	At 50° F.
Benzine	0.883	At 59° F.
Chloroform	1.491	At 62.6° F.
Carbon bisulphide	1.293 1	At 32° F.
Ether	0.720 4	At 60.8° F.
Glycerine	1.263 6	At 59° F.
Hydrochloric acid	1.27	
Mercury	13.596	At 32° F.
Nitric acid	1.552	At 59° F.
Oil of turpentine	0.855 to 0.864	At 68° F.
Linseed oil	0.94	
Olive oil	0.915	
Sulphuric acid	1.854	At 32° F.

Gases.

Names of gases.	At 0° C. and 760 mm. pressure compared to water.	At 0° C. and 760 mm. pressure compared to air.
Air	0.001 292 8	1.
Oxygen	0.001 429 3	1.105 63
Nitrogen	0.001 255 7	0.971 37
Hydrogen	0.000 089 54	0.069 26
Carbonic dioxide	0.001 976 7	1.529 1
Mixed gases from electrolysis of water	0.000 536 1	0.414 72
Aqueous vapor		0.623

WEIGHTS OF SUBSTANCES.

Names of substances.	Average weights per cubic foot. Pounds.
Asphaltum	87.
Brick, common, hard	125.
Brickwork, pressed brick	140.
" ordinary	112.
Coal, anthracite, solid, of Pennsylvania	93.
" " broken, loose	54.
" bituminous, solid	84.
" " broken, loose	49.
Coke, loose, of good coal	62.
Cork	12.4
Earth, common loam, dry, loose	76.
" " " " moderately rammed	95.
" as a soft, flowing mud	108.
Gneiss, common	168.
Granite	170.
Glass, Crown	168.5
" flint	218.3
Ice at 0° C	57.2
Lime, thoroughly shaken	75.
Masonry, of granite or limestone, well dressed	165.
Mortar, hardened	103.
Mud, dry, close	80 to 1
Quartz	165.4
Sulphur	131.
Wax	58.7
Wood, ebony	74.9
" birch	43.7
" oak	46.8
" pine	31.2
Water at 32° F	62.418
" " 39.1° F	62.425
" " 50° F	62.409
" " 60° F	62.367
" " 70° F	62.302
" " 80° F	62.218
" " 90° F	62.119

THE COMPARISON OF THERMOMETERS.

Fahrenheit to Centigrade.

($t°$ F. $- 32$) $\times \frac{5}{9} =$ Degrees C.

Fahrenheit	Centigrade	Fahrenheit	Centigrade	Fahrenheit	Centigrade	Fahrenheit	Centigrade	Fahrenheit	Centigrade
50	10.	61	16.1	72	22.2	83	28.3	94	34.4
51	10.6	62	16.7	73	22.8	84	28.9	95	35.
52	11.1	63	17.2	74	23.3	85	29.4	96	35.6
53	11.7	64	17.8	75	23.9	86	30.	97	36.1
54	12.2	65	18.3	76	24.4	87	30.6	98	36.7
55	12.8	66	18.9	77	25.	88	31.1	99	37.2
56	13.3	67	19.4	78	25.6	89	31.7	100	37.8
57	13.9	68	20.	79	26.1	90	32.2		
58	14.4	69	20.6	80	26.7	91	32.8		
59	15.	70	21.1	81	27.2	92	33.3		
60	15.6	71	21.7	82	27.8	93	33.9		

Centigrade to Fahrenheit.

$\frac{9}{5} t°$ C $+ 32 =$ Degrees F.

Centigrade	Fahrenheit	Centigrade	Fahrenheit	Centigrade	Fahrenheit	Centigrade	Fahrenheit
10	50.	18	64.4	26	78.8	34	93.2
11	51.8	19	66.2	27	80.6	35	95.
12	53.6	20	68.	28	82.4	36	96.8
13	55.4	21	69.8	29	84.2	37	98.6
14	57.2	22	71.6	30	86.	38	100.4
15	59.	23	73.4	31	87.8	39	102.2
16	60.8	24	75.2	32	89.6	40	104.
17	62.6	25	77.	33	91.4		

ELECTRICAL UNITS.

Final and official recommendation of the Chamber of Delegates of the International Electrical Congress, held at Chicago, 1893.

Resolved, That the several governments represented by the delegates of this International Congress of Electricians be, and they are hereby, recommended to formally adopt as legal units of electrical measure the following: As a unit of resistance, the *international ohm*, which is based upon the ohm equal to 10^9 units of resistance of the C. G. S. system of electro-magnetic units, and is represented by the resistance offered to an unvarying electric current by a column of mercury at the temperature of melting ice 14.452 1 grams in mass, of a constant cross-sectional area and of the length of 106.3 centimeters.

As a unit of current, the *international ampere*, which is one-tenth of the unit of current of the C. G. S. system of electro-magnetic units, and which is represented sufficiently well for practical use by the unvarying current which, when passed through a solution of nitrate of silver in water, and in accordance with accompanying specifications,[1] deposits silver at the rate of 0.001 118 of a gram per second.

---

1. In the following specification the term silver voltameter means the arrangement of apparatus by means of which an electric current is passed through a solution of nitrate of silver in water. The silver voltameter measures the total electrical quantity which has passed during the time of the experiment, and by noting this time the time average of the current, or, if the current has been kept constant, the current itself can be deduced.

In employing the silver voltameter to measure currents of about one ampere, the following arrangements should be adopted:

The kathode on which the silver is to be deposited should take the form of a platinum bowl not less than 10 centimeters in diameter and from 4 to 5 centimeters in depth.

The anode should be a plate of pure silver, some 30 square centimeters in area and 2 or 3 millimeters in thickness.

This is supported horizontally in the liquid near the top of the solution by a platinum wire passed through holes in the plate at opposite corners. To prevent the disintegrated silver which is formed on the anode from falling onto the kathode, the anode should be wrapped around with pure filter paper, secured at the back with sealing wax.

The liquid should consist of a neutral solution of pure silver nitrate, containing about 15 parts by weight of the nitrate to 85 parts of water.

The resistance of the voltameter changes somewhat as the current

As a unit of electro-motive force, the *international volt*, which is the electro-motive force that, steadily applied to a conductor whose resistance is one international ohm, will produce a current of one international ampere, and which is represented sufficiently well for practical use by $\frac{1000}{1434}$ of the electro-motive force between the poles or electrodes of the voltaic cell known as Clark's cell, at a temperature of 15° C., and prepared in the manner described in the accompanying specification.[2]

As a unit of quantity, the *international coulomb*, which is the quantity of electricity transferred by a current of one international ampere in one second.

As a unit of capacity, the *international farad*, which is the capacity of a condenser charged to a potential of one international volt by one international coulomb of electricity.

As a unit of work, the *joule*, which is equal to 10^7 units of work in the C. G. S. system, and which is represented sufficiently well for practical use by the energy expended in one second by an international ampere in an international ohm.

As a unit of power, the *watt*, which is equal to 10^7 units of power in the C. G. S. system, and which is represented sufficiently well for practical use by work done at the rate of one joule per second.

As the unit of induction, the *henry*, which is the induction in a circuit when the electro-motive force induced in this circuit is one international volt, while the inducing current varies at the rate of one ampere per second.

---

passes. To prevent these changes having too great an effect on the current, some resistance besides that of the voltameter should be inserted in the circuit. The total metallic resistance of the circuit should not be less than 10 ohms.

2. A committee, consisting of Messrs. Helmholtz, Ayrton and Carhart, was appointed to prepare specifications for the Clark's cell. Their report has not yet been received.

COPPER WIRE.

IN THE following tables of copper wire the value of the mil-foot is taken as the standard.

The temperature coëfficient is interpolated for 60° F. and 75° F. from the values given in the second table.

In the table for B. & S. G., the actual sizes to which wire is drawn, are used.

In many cases the nearest whole number of pounds is taken when the variation is less than that found in actual weights of drawn wire.

In computing the weights, the specific gravity of copper is taken at 8.89, water being at its greatest density 62.425 pounds per cubic foot.

International ohms are used, unless the kind of unit is specifically stated.

The following formulæ were used:

$$\text{Resistance per 1 000 feet at } 60° \text{ F.} = \frac{10\,180.694}{d^2}$$

$$\text{Resistance per 1 000 feet at } 75° \text{ F.} = \frac{10\,507.4}{d^2}$$

Weight per 1 000 feet = $.003\,027 \times d^2$.

Weight per mile = $.015\,983 \times d^2$.

The following data and formulæ may be useful:

One B. A. unit = .988 9 legal ohms = .986 6 International ohms.
One legal ohm = 1.011 22 B. A. units = .997 67 International ohms.
One International ohm = 1.013 58 B. A. units = 1.002 33 legal ohms.
One cubic foot of copper weighs 555 pounds.
One cubic inch of copper weighs .321 2 pounds.

$$\text{Resistance per 1 000 feet at } 60° \text{ F.} = \frac{30.815}{\text{weight per 1 000 feet}}$$

$$\text{Resistance per 1 000 feet at } 75° \text{ F.} = \frac{31.804}{\text{weight per 1 000 feet}}$$

If a copper wire of length l, diameter d, and weight w, has a resistance R at temperature t, then its conductivity by diameter is given by the first formula, and by weight by the second.

$$C = \frac{a\,l\,k}{d^2\,R} \qquad R\,t° = \frac{a\,l\,k}{d^2}$$

$$C = \frac{b\,l^2\,c}{w\,R} \qquad R\,t° = \frac{b\,l^2\,c}{w}$$

Here, a is the resistance of a mil-foot in same units as R, k is the temperature coëfficient for t° Centigrade, and b is the resistance of one meter-gram at temperature t° and in same units as R.

When l is in meters and w in grams, $c = 1$.
When l is in feet and w in grams, $c = .092\,9$.
When l is in feet and w in pounds, $c = .000\,204\,8$.

Mile-ohm = weight per mile × resistance per mile.

Mile-ohm at 60° = 859, International ohms.
Mile-ohm at 60° = 868.9, B. A. units.
Mile-ohm at 60° = 861, Legal ohms.

The following tables are taken from the report of the Standard Wiring Table Committee, published in the report of the meeting of the American Institute of Electrical Engineers, held January 17, 1893:

MATTHIESSEN'S STANDARD.
(Recommended by the Committee).

Equivalent length of a square mm. mercury column.	B. A. units. 104.8 cms.	Legal ohms. 106.0 cms.	International ohms. 106.3 cms.
Resistance at 0° C. of Matthiessen's Standard—			
Meter-gram soft copper............	.143 65	.142 06	.141 73
Meter-millimeter soft copper...........................	.020 57	.020 35	.020 3
Cubic centimeter soft copper...........................	.000 001 616	.000 001 598	.000 001 594
Mil-foot soft copper............	9.72	9.612	9.59

TEMPERATURE COËFFICIENTS.

Table of temperature variations in the resistance of pure soft copper according to Matthiessen's standard and formulæ.

Temperature in degrees Centigrade.	Temperature coëfficient of resistance.	Logarithm.	Matthiessen meter-gram standard resistance.		
			B. A. units.	Legal ohms.	International ohms.
0	1.	0.	0.143 65	0.142 06	0.141 73
1	1.003 876	0.001 680 1	0.144 21	0.142 61	0.142 28
2	1.007 764	0.003 358 8	0.144 77	0.143 17	0.142 83
3	1.011 66	0.005 036 2	0.145 33	0.143 72	0.143 38
4	1.015 58	0.006 712 1	0.145 89	0.144 27	0.143 94
5	1.019 5	0.008 386 4	0.146 45	0.144 83	0.144 49
6	1.023 43	0.010 059 3	0.147 02	0.145 39	0.145 05
7	1.027 38	0.011 730 7	0.147 59	0.145 95	0.145 61
8	1.031 34	0.013 400 3	0.148 15	0.146 51	0.146 17
9	1.035 31	0.015 068 3	0.148 73	0.147 08	0.146 73
10	1.039 29	0.016 734 6	0.149 3	0.147 64	0.147 3
11	1.043 28	0.018 399 3	0 149 87	0.148 21	0.147 86
12	1.047 28	0.020 062 1	0.150 45	0.148 78	0.148 43
13	1.051 29	0.021 723	0.151 02	0.149 35	0.149
14	1.055 32	0.023 382 1	0.151 6	0.149 92	0.149 57
15	1.059 35	0.025 039	0.152 18	0.150 49	0.150 14
16	1.063 39	0.026 694	0.152 77	0.151 07	0.150 71
17	1.067 45	0.028 348	0.153 34	0.151 64	0.151 29
18	1.071 52	0.029 999	0.153 93	0.152 22	0.151 86
19	1.075 59	0.803 164	0.154 51	0.152 8	0.152 44
20	1.079 68	0.033 294	0.155 1	0.153 38	0.153 02
21	1.083 78	0.034 939	0.155 69	0.153 96	0.153 6
22	1.087 88	0.036 581	0.156 28	0.154 55	0.154 18
23	1.092	0.038 222	0.156 87	0.155 13	0.154 77
24	1 096 12	0.039 859	0.157 46	0.155 72	0.155 35
25	1.100 26	0.041 494	0.158 06	0.156 31	0.155 94
26	1.104 4	0.043 127	0.158 65	0.156 89	0.156 53
27	1.108 56	0.044 758	0.159 25	0.157 48	0.157 11
28	1.112 72	0.046 385	0.159 85	0 158 08	0.157 7
29	1.116 89	0.048 011	0.160 44	0.158 67	0.158 3
30	1.121 07	0.049 633	0.161 05	0.159 26	0.158 89
40	1.163 32	0.065 699	0.167 11	0.165 26	0.164 88
50	1.206 25	0.081 436	0.173 28	0.171 36	0.170 96
60	1.249 65	0.096 787	0.179 52	0.177 53	0.177 11
70	1.293 27	0.111 687	0.185 78	0.183 72	0.183 29
80	1.336 81	0.126 069	0.192 04	0.189 91	0.189 46
90	1.379 95	0.139 863	0.198 23	0.196 04	0.195 58
100	1 422 31	0.152 995	0.204 32	0.202 06	0.201 58

PROPERTIES OF COPPER WIRE.

English system—Brown & Sharpe gauge.

Number.	Diameters in mils.	Areas in circular mils. $C.M. = d^2$.	Weights.		Resistances per 1 000 feet in International ohms.	
			1 000 feet.	Mile.	At 60° F.	At 75° F.
0 000	460.	211 600.	641.	3 382.	.048 11	.049 66
000	410.	168 100.	509.	2 687.	.060 56	.062 51
00	365.	133 225.	408.	2 129.	.076 42	.078 87
0	325.	105 625.	320.	1 688.	.096 39	.099 48
1	289.	83 521.	253.	1 335.	.121 9	.125 8
2	258.	66 564.	202.	1 064.	.152 9	.157 9
3	229.	52 441.	159.	838.	.194 1	.200 4
4	204.	41 616.	126.	665.	.244 6	.252 5
5	182.	33 124.	100.	529.	.307 4	.317 2
6	162.	26 244.	79.	419.	.387 9	.400 4
7	144.	20 736.	63.	331.	.491	.506 7
8	128.	16 384.	50.	262.	.621 4	.641 3
9	114.	12 996.	39.	208.	.783 4	.808 5
10	102.	10 404.	32.	166.	.978 5	1.01
11	91.	8 281.	25.	132.	1.229	1.269
12	81.	6 561.	20.	105.	1.552	1.601
13	72.	5 184.	15.7	83.	1.964	2.027
14	64.	4 096.	12.4	65.	2.485	2.565
15	57.	3 249.	9.8	52.	3.133	3.234
16	51.	2 601.	7.9	42.	3.914	4.04
17	45.	2 025.	6.1	32.	5.028	5.189
18	40.	1 600.	4.8	25.6	6.363	6.567
19	36.	1 296.	3.9	20.7	7.855	8.108
20	32.	1 024.	3.1	16.4	9.942	10.26
21	28.5	812.3	2.5	13.	12.53	12.94
22	25.3	640.1	1.9	10.2	15.9	16.41
23	22.6	510.8	1.5	8.2	19.93	20.57
24	20.1	404.	1.2	6 5	25.2	26.01
25	17.9	320.4	.97	5.1	31.77	32.79
26	15.9	252.8	.77	4.	40.27	41.56
27	14.2	201.6	.61	3.2	50.49	52.11
28	12.6	158.8	.48	2.5	64.13	66.18
29	11.3	127.7	.39	2.	79.78	82.29
30	10.	100.	.3	1.6	101.8	105.1
31	8.9	79.2	.24	1.27	128.5	132.7
32	8.	64	.19	1.02	159.1	164.2
33	7.1	50.4	.15	.81	202.	208.4
34	6.3	39.7	.12	.63	256.5	264.7
35	5.6	31.4	.095	.5	324.6	335.1
36	5.	25.	.076	.4	407.2	420.3

PROPERTIES OF COPPER WIRE.—(Cont.)
English system—Birmingham wire gauge.

Numbers.	Diameters in mils.	Areas in circular mils. $C. M. = d^2$	Weights. 1 000 feet.	Weights. Mile.	Resistances per 1 000 feet in International ohms. At 60° F.	Resistances per 1 000 feet in International ohms. At 75° F.
0 000	454.	206 116.	624.	3 294.	.049 39	.050 98
000	425.	180 625.	547.	2 887.	.056 36	.058 17
00	380.	144 400.	437.	2 308.	.070 5	.072 77
0	340.	115 660.	350.	1 847.	.088 07	.090 89
1	300.	90 000.	272.	1 438.	.113 1	.116 7
2	284.	80 656.	244.	1 289.	.126 2	.130 3
3	259.	67 081.	203.	1 072.	.151 8	.156 6
4	238.	56 644.	171.	905.	.179 7	.185 5
5	220.	48 400.	146.	773.	.210 3	.217 1
6	203.	41 209.	125.	659.	.247 1	.255
7	180.	32 400.	98.	518.	.314 2	.324 3
8	165.	27 225.	82.	435.	.373 9	.385 9
9	148.	21 904.	66.	350.	.464 8	.479 7
10	134.	17 956.	54.	287.	.567	.585 2
11	120.	14 400.	44.	230.	.707	.729 7
12	109.	11 881.	36.	190.	.856 9	.884 4
13	95.	9 025.	27.3	144.	1.128	1.164
14	83.	6 889.	20.8	110.	1.478	1.525
15	72.	5 184.	15.7	83.	1.964	2.027
16	65.	4 225.	12.8	68.	2.41	2 487
17	58.	3 364.	10.2	54	3.026	3.123
18	49.	2 401.	7.3	38.4	4.24	4.376
19	42.	1 764.	5 3	28.2	5.771	5.957
20	35.	1 225.	3.7	19.6	8.311	8.577
21	32.	1 024.	3.1	16.4	9.942	10 26
22	28.	784.	2.4	12.5	12.99	13.4
23	25.	625.	1.9	10.	16.29	16.81
24	22.	484.	1.5	7.7	21.08	21.71
25	20.	400.	1.2	6.4	25.45	26.27
26	18.	324.	.98	5 2	31.42	32.43
27	16.	256.	.77	4.1	39.77	41.04
28	14.	196.	.59	3.1	51.94	53.61
29	13.	169.	.51	2.7	60.24	62.17
30	12.	144.	.44	2.3	70.7	72.97
31	10.	100.	.3	1 6	108.	105.1
32	9.	81.	.25	1.3	125.7	129.7
33	8.	64.	.19	1.02	159.1	164.2
34	7.	49.	.15	.78	207.8	214.4
35	5.	25.	.075	.4	407.2	420.3
36	4.	16.	.048	.256	636.3	656.7

PROPERTIES OF COPPER WIRE.—(Cont.)
English system—New British standard gauge.

Numbers.	Diameters in mils.	Areas in Circular mils. C M. = d².	Weights. 1 000 feet.	Weights. Mile.	Resistances per 1 000 feet in International ohms. At 60° F.	Resistances per 1 000 feet in International ohms. At 75° F.
000 000	464.	215 296.	652.	3 441.	.047 29	.048 8
00 000	432.	186 624.	565.	2 983.	.054 55	.056 3
0 000	400.	160 000.	484.	2 557.	.063 63	.065 67
000	372.	138 384.	419.	2 212.	.073 57	.075 93
00	348.	121 104.	367.	1 935.	.084 07	.086 76
0	324.	104 976.	818.	1 678.	.969 8	.100 09
1	300.	90 000.	272.	1 438.	.113 1	.116 7
2	276.	76 176.	231.	1 217.	.133 6	.137 9
3	252.	63 504.	192.	1 015.	.160 3	.165 5
4	232.	53 824.	163.	860.	.189 2	.195 2
5	212.	44 914.	136.	718.	.226 5	.233 8
6	192.	36 864.	112.	589.	.276 2	.285
7	176.	30 976.	94.	495.	.328 7	.339 2
8	160.	25 600.	77.	409.	.397 7	.410 4
9	144.	20 736.	63.	331.	.491	.506 7
10	128.	16 384.	50.	262.	.621 4	.641 3
11	116.	13 456.	41.	215.	.756 6	.780 9
12	104.	10 816.	83.	173.	.941 3	.971 5
13	92.	8 464.	25.6	135.	1.203	1.241
14	80.	6 400.	19.4	102.	1.591	1.642
15	72.	5 184.	15.7	83.	1.964	2.027
16	64.	4 096.	12.4	65.	2.486	2.565
17	56.	3 136.	9.5	50.	3.246	3.351
18	48.	2 304.	7.	36.8	4.419	4.561
19	40.	1 600.	4.8	25.6	6.363	6.567
20	36.	1 296.	3.9	20.7	7.855	8.108
21	32.	1 024.	3.1	16.4	9.942	10.26
22	28.	784.	2.4	12.5	12.99	13.4
23	24.	576.	1.7	9.2	17.67	18.24
24	22.	484.	1.5	7.7	21.03	21.71
25	20.	400.	1.2	6.4	25.45	26.27
26	18.	324.	.98	5.2	31.42	32.43
27	16.4	269.	.81	4.3	37.85	39.07
28	14.8	219.	.66	3.5	46.48	47.97
29	13.6	185.	.56	3.	55.04	56.81
30	12.4	153.8	.47	2.5	66.21	68.34
31	11.6	134.6	.41	2.15	75.66	78.09
32	10.8	116.6	.35	1.86	87.28	90.08
33	10.	100.	.3	1.6	101.8	105.1
34	9.2	84.6	.26	1.35	120.3	124.1
35	8.4	70.6	.21	1.13	144.3	148.9
36	7.6	57.8	.17	.92	176.3	181.9

PROPERTIES OF COPPER WIRE.—(*Cont.*)
Metric system—Brown & Sharpe gauge.

Numbers	Diameters in millimeters.	Areas in square millimeters.	Weights per kilometer in kilograms.	Resistances per kilometer in International ohms.	
				At 60° F.	At 75° F.
0 000	11.683	107.2	954.3	.157 8	.162 9
000	10.404	85.01	756.8	.198 7	.205 1
00	9.266	67.43	600.2	.250 7	.258 8
0	8.251	53.47	480.4	.316 2	.326 4
1	7.348	42.41	377.4	.399 9	.412 7
2	6.544	33.63	299.3	.501 8	.517 9
3	5.827	26.67	237.4	.636 9	.657 4
4	5.19	21.16	188.3	.802 6	.828 4
5	4.621	16.77	149.3	1.009	1.041
6	4.115	13.3	118.4	1.273	1.314
7	3.665	10.55	93.9	1.611	1.662
8	3.263	8.362	74.5	2.039	2.104
9	2.906	6.633	59.	2.57	2.653
10	2.588	5.26	46.8	3.21	3.313
11	2.305	4.173	37.1	4.033	4.163
12	2.052	3.307	29.5	5.091	5.253
13	1.828	2.625	23.4	6.443	6.65
14	1.628	2.082	18.5	8.155	8.416
15	1.449	1.649	14.7	10.28	10.61
16	1.291	1.309	11.7	12.84	13.25
17	1.15	1.039	9.23	16.5	17.02
18	1.024	.823 6	7.32	20.88	21.55
19	.911 6	.652 7	5.8	25.77	26.6
20	.811 8	.517 6	4.61	32.62	33.66
21	.722 9	.410 4	3.65	41.11	42.45
22	.643 8	.325 5	2.89	52.16	53.84
23	.573 3	.258 1	2.16	65.39	67.49
24	.510 5	.204 7	1.82	82.68	85.33
25	.454 6	.162 3	1.44	104.2	107.6
26	.404 9	.128 8	1.15	132.1	136.3
27	.360 5	.102 1	.908	165.1	171.
28	.321 1	.081	.72	210.4	217.1
29	.285 9	.064 2	.572	261.6	270.
30	.254 5	.050 9	.452	334.	344.8
31	.226 7	.040 4	.359	421.6	435.4
32	.201 9	.032	.284	522.	538.7
33	.179 8	.025 4	.226	662.7	683.7
34	.160 1	.020 1	.179	841.5	868.4
35	.142 6	.016	.141	1 065.	1 099.
36	.127	.012 7	.113	1 336.	1 379.

WEIGHTS OF COPPER WIRE.

Metric system—per kilometer, in kilograms.

Numbers.	Roebling.	Brown & Sharpe.	Birmingham or Stubs.	New British standard.
000 000	954.3	....	....	970.9
00 000	833.9	....	....	841.6
0 000	696.5	954.3	929.4	721.5
000	591.	756.8	814.5	624.
00	494.1	600.2	651.3	546.2
0	425.1	480.4	521.3	473.4
1	361.2	377.4	405.8	405.8
2	311.9	299.3	363.3	343.5
3	268.5	237.4	302.6	286.3
4	228.3	188.3	255.3	.242.7
5	193.2	149.3	218.3	202.7
6	166.2	118.4	185.9	166.2
7	141.3	93.9	146.1	139.7
8	118.3	74.5	122.8	115.4
9	98.8	59.	98.8	93.5
10	82.2	46.8	81.	73.9
11	64.9	37.1	64.9	60.7
12	49.9	29.5	53.6	48.8
13	38.2	23.4	39.8	38.2
14	28.9	18.5	31.1	28.9
15	23.4	14.7	23.4	23.4
16	17.9	11.7	19.1	18.5
17	13.2	9.23	15.2	14.1
18	9.96	7.32	10.8	10.4
19	7.58	5.8	7.95	7.22
20	5.52	4.61	5.52	5.85
21	4.61	3.65	4.62	4.61
22	3.54	2.89	3.54	3.54
23	2.81	2.16	2.81	2.59
24	2.38	1.82	2.19	2.19
25	1.8	1.44	1.8	1.8
26	1.46	1.15	1.46	1.46
27	1.3	.908	1.16	1.21
28	1.15	.72	.884	.988
29	1.02	.572	.762	.833
30	.884	.452	.649	.694
31	.822	.359	.451	.607
32	.762	.284	.365	.525
33	.544	.226	.289	.451
34	.451	.179	.22	.381
35	.406	.141	.113	.319
36	.365	.113	.071	.26

HARD-DRAWN COPPER WIRE.

British Post-office specifications.

Diameters.			Weights per mile.			Minimum breaking strain. Pounds.	Minimum twists.		Maximum resistance per mile at 60° F. International ohms.
Required.	Maximum.	Minimum.	Required.	Maximum.	Minimum.				
224	226	220½	800	820	780	2400	in 6 in.	15	1.098
194	196	191	600	615	585	1800		20	1.464
158	160¼	155½	400	410	390	1300		25	2.195
112	113¼	110½	200	205	195	650	in 3 in.	20	4.391
97	98	95½	160	153¾	146¼	490		25	5.855
79	80	78	100	102½	97½	330		30	8.782

"The wire shall be capable of being wrapped in six turns around wire of its own diameter, unwrapped and again wrapped in six turns around wire of its own diameter in the same direction as the first wrapping, without breaking; and shall be also capable of bearing the number of twists set down in the table, without breaking.

"The twist-test will be made as follows: The wire will be gripped by two vises, one of which will be made to revolve at a speed not exceeding one revolution per second. The twists thus given to the wire will be reckoned by means of an ink mark which forms a spiral on the wire during torsion, the full number of twists to be visible between the vises."

According to the above table, the mile-ohm of copper required is 878 pounds. This corresponds to a conductivity of 96.6 per cent., taking the value of the mile-ohm of 100 per cent. copper as 859.

HARD-DRAWN COPPER WIRE.—(Continued.)
Telephone specifications.

Numbers.	Diameters in mils.			Weights per mile.			Breaking weights.			Weights of coils.		Conductivity.		Twists in six inches.	Per cent. elongation in five feet.
	Required.	Maximum.	Minimum.	Required.	Maximum.	Minimum.	Actual required.	Actual minimum.	Per square inch.	Maximum.	Minimum.	Required.	Minimum.		
8 B. W. G......	165.	166.	164.	436.4	441.7	431.1	1 328	1 301	62 100	218	152	97	96	30	1.14
12 N. B. S. G...	104.	104.9	103.1	173.4	176.4	170.4	549	538	64 600	219	151	97	96	40	1.
10 B. & S. G...	101.9	102.8	101.	165.	168.	162.	540	519	64 800	218	152	97	96	40	.99
12 B. & S. G...	80.	81.2	79.3	102.6	105.7	100.8	334	327	66 500	72	62	97	96	44	.94
14 B. & S. G...	64.	65.	63.	65.	67.5	63.	220	212	68 200	....	....	97	96	47	.91

TENSILE STRENGTH OF COPPER WIRE.

Numbers, B. & S. G.	Breaking weight. Pounds.		Numbers, B. & S. G.	Breaking weight. Pounds.	
	Hard-drawn.	Annealed.		Hard-drawn.	Annealed.
0 000	8 310	5 650	9	617	349
000	6 580	4 480	10	489	277
00	5 226	3 553	11	388	219
0	4 558	2 818	12	307	174
1	3 746	2 234	13	244	138
2	3 127	1 772	14	193	109
3	2 480	1 405	15	153	87
4	1 967	1 114	16	133	69
5	1 559	883	17	97	55
6	1 237	700	18	77	43
7	980	555	19	61	34
8	778	440	20	48	27

The strength of soft copper wire varies from 32 000 to 36 000 pounds per square inch, and of hard copper wire from 45 000 to 68 000 pounds per square inch, according to the degree of hardness.

The above table is calculated for 34 000 pounds for soft wire and 60 000 pounds for hard wire, except for some of the larger sizes, where the breaking weight per square inch is taken at 50 000 pounds for 0 000, 000 and 00, 55 000 for 0, and 57 000 pounds for 1.

BI-METALLIC WIRE.

Numbers, B. & S. G.	Diameters in mils.	Weights per mile. Pounds.	Breaking weight. Pounds.
0 000	460	3 200	10 500
000	410	2 537	8 600
00	365	2 022	7 000
0	325	1 620	5 700
1	289	1 264	4 600
2	258	1 008	3 800
3	229	797	3 200
4	204	629	2 600
5	182	490	1 790
6	162	398	1 500
7	144	314	1 210
8	128	246	1 020
9	114	203	850
10	102	157	660
11	91	127	520
12	81	100	410
14	64	63	260
16	51	40	160
18	40	25	100

This wire consists of a steel center with a cover of copper. Its conductivity is about 65 per cent. of that of pure copper. The percentage of copper and steel may vary a trifle, hence the strength and weight must be approximate.

STRANDS OF COPPER WIRE.

COPPER WIRES are laid up into concentric strands or into ropes of seven strands. A rope of seven strands each composed of seven wires, is called a seven by seven rope, and is usually written 7x7. The number of wires that can be made into a strand is limited by the capacity of the stranding machinery. Two hundred wires is the usual limit of a concentric strand, and one hundred and thirty-three wires of a rope.

In a strand of circular milage, C. M., composed of n wires of diameter d, with a weight per 1 000 feet w, then we have

$$C. M. = d^2 \times n.$$

$$n = \frac{C. M.}{d^2}$$

$$d = \sqrt{\frac{C. M.}{n}}$$

$$w = .00305 \times C. M.$$

The weights of strands are calculated about one per cent. heavier than a solid wire of the same circular milage, while the resistance is calculated for the solid wire.

In specifying how a strand shall be made, the number of wires to be used or the diameter of each wire may be given. In the first case the wire usually has to be specially drawn, and this will delay an order, especially a small order, unduly. It is, therefore, better to specify the size wires B. & S. G., of which the strand is to be made.

The diameter of a strand may be calculated by multiplying the diameter of one wire by the factors given in the table at the bottom of the opposite page, according to the number of wires composing the strand.

STRANDS OF COPPER WIRE.
Diameters and properties.

Numbers, B. & S. G.	Circular mils.	Diameters.		Weights.		Resistances at 75° F. per 1 000 ft.
		Decimal parts of inch.	Nearest 32d.	1 000 feet.	Mile.	
....	1 000 000	1.152	1 5/32	3 050	16 104	.010 51
....	950 000	1.125	1 1/8	2 898	15 299	.011 06
....	900 000	1.092	1 3/32	2 745	14 494	.011 67
....	850 000	1.062	1 2/32	2 593	13 688	.012 36
....	800 000	1.035	1 1/32	2 440	12 883	.013 13
....	750 000	.999	1	2 288	12 078	.014 01
....	700 000	.963	31/32	2 135	11 273	.015 01
....	650 000	.927	15/16	1 983	10 468	.016 17
....	600 000	.891	29/32	1 830	9 662	.017 51
....	550 000	.855	7/8	1 678	8 857	.019 1
....	500 000	.819	13/16	1 525	8 052	.021 01
....	450 000	.770	25/32	1 373	7 247	.023 35
....	400 000	.728	3/4	1 220	6 442	.026 27
....	350 000	.679	11/16	1 068	5 636	.030 02
....	300 000	.630	5/8	915	4 831	.035 02
	250 000	.590	19/32	762	4 026	.042 03
0 000	211 600	.530	17/32	645	3 405	.049 66
000	168 100	.470	15/32	513	2 709	.062 51
00	133 225	.420	7/16	406	2 144	.078 87
0	105 625	.375	3/8	322	1 700	.099 48
1	83 521	.330	11/32	255	1 346	.125 8
2	66 564	.291	9/32	203	1 072	.157 9
3	52 441	.261	17/64	160	845	.200 4
4	41 616	.231	1/4	127	671	.252 5

Numbers of wires.	Factors.	Numbers of wires.	Factors.
3	2 1/4	75	10 1/4
7	3	91	11
12	4 1/4	108	12 1/4
19	5	127	13
27	6 1/4	147	14 1/4
37	7	169	15
48	8 1/4	192	16 1/4
61	9	217	17
7x7	9	.....	......
7x19	15	.....	......

DIAMETERS OF WIRES IN STRANDS.

Circular mils.	Numbers of wires.														
	1	7	12	19	27	37	48	61	75	91	108	127	147	169	192	217	7x7	7x19
	Diameter of each wire.															
1 000 000	1000.	377.	288.7	229.4	192.5	164.4	144.8	128.	115.5	104.8	96.2	88.7	82.5	76.9	72.2	67.8	142.9	86.7
950 000	974.6	368.4	281.4	223.6	187.6	160.2	140.7	124.7	112.6	102.1	93.8	86.4	80.4	74.9	70.8	66.1	139.2	84.5
900 000	948.6	358.5	273.9	217.6	182.6	155.9	136.9	121.4	109.5	99.5	91.3	84.1	78.3	72.9	68.5	64.4	135.5	82.2
850 000	921.9	348.4	266.1	211.5	177.4	151.5	133.1	118.	106.5	96.8	88.7	81.8	76.	70.9	66.5	62.5	131.7	79.9
800 000	894.4	338.	258.2	205.	172.1	147.	129.1	114.5	103.8	93.7	86.1	79.3	73.8	68.7	64.5	60.7	127.8	77.5
750 000	866.	327.3	250.	198.6	166.7	142.3	125.	110.8	100.	90.7	83.3	76.8	71.4	66.6	62.5	58.8	123.7	75.
700 000	836.6	316.3	241.5	191.9	161.	137.5	120.8	107.1	96.6	87.7	80.5	74.2	69.	64.3	60.4	56.7	119.5	72.5
650 000	806.2	304.7	232.7	184.9	155.2	132.5	116.4	103.2	93.1	84.5	77.6	71.5	66.5	62.	58.2	54.7	115.2	69.9
600 000	774.6	292.7	223.6	177.6	149.1	127.3	111.8	99.1	89.4	81.2	74.5	68.7	63.9	59.5	55.9	52.5	110.7	67.1
550 000	741.6	280.3	214.1	170.1	142.7	121.9	107.	94.9	85.6	77.7	71.4	65.8	61.2	57.1	53.5	50.3	106.	64.2
500 000	707.1	267.2	204.1	162.2	138.1	116.2	102.1	90.5	81.7	74.1	68.	62.7	58.3	54.3	51.	48.	101.	61.8
450 000	670.8	253.5	193.7	153.8	129.1	110.3	96.8	85.8	77.5	70.3	64.6	59.5	55.3	51.6	48.4	45.5	95.8	58.1
400 000	632.4	239.	182.6	145.	121.7	103.9	91.3	80.9	73.	66.3	60.9	56.1	52.2	48.6	45.6	42.9	90.4	54.8
350 000	591.6	223.6	170.8	135.7	113.9	97.2	85.4	75.7	68.3	62.	56.9	52.6	48.8	45.5	42.7	40.1	84.5	51.2
300 000	547.7	207.	158.1	125.6	105.4	90.	79.1	70.1	63.2	57.4	52.7	48.6	45.2	42.1	39.5	37.1	78.3	47.4
250 000	500.	189.	144.8	114.7	96.2	82.1	72.2	64.	57.7	52.4	48.1	44.3	41.2	38.4	36.1	33.9	71.4	43.8

DIAMETERS OF WIRES IN STRANDS.—(Continued.)

Numbers, B. & S. G.	3	7	12	19	27	37	48	61	75	91	108	127	147	169	192	217	7x7	7x19
								Diameter of each wire.										
000	265.6	173.9	132.8	105.5	88.5	75.6	67.9	58.9	53.1	48.2	44.3	40.8	37.9	35.4	33.2	31.2	65.7	39.9
000	236.7	155.	118.8	94.1	78.9	67.4	59.2	52.5	47.8	43.	39.5	36.4	33.8	31.5	29.6	27.8	58.6	35.6
00	210.7	138.	105.4	83.7	70.8	60.	52.7	46.7	42.2	38.3	35.1	32.4	30.1	28.1	26.3	24.8	52.1	31.7
0	187.7	122.8	93.8	74.6	62.6	53.4	46.9	41.6	37.5	34.1	31.3	28.8	26.8	25.	23.5	22.1	46.4	28.2
1	166.9	109.2	83.4	66.3	55.6	47.5	41.7	37.	33.4	30.8	27.8	25.6	23.8	22.2	20.9	19.6	41.3	25.1
2	149.	97.5	74.5	59.2	49.7	42.4	37.2	33.	29.8	27.1	24.8	22.9	21.3	19.9	18.6	17.5	36.9	22.4
3	132.2	86.6	66.1	52.5	44.1	37.7	33.1	29.3	26.5	24.	22.	20.3	18.9	17.6	16.5	15.6	32.7	19.9
4	117.8	77.1	58.9	46.8	39.3	33.5	29.5	26.1	23.6	21.4	19.6	18.1	16.8	15.7	14.7	13.9	29.1	17.7
5	105.1	68.8	52.5	41.8	35.	29.9	26.3	23.3	21.	19.1	17.5	16.2	15.	14.	13.1	12.1	26.	15.8
6	93.5	61.2	46.8	37.2	31.2	26.6	23.4	20.7	18.7	17.	15.6	14.4	13.4	12.5	11.7	11.	23.1	14.1
8	73.9	48.4	37.	29.4	24.6	21.1	18.5	16.4	14.8	13.4	12.8	11.4	10.6	9.8	9.2	8.7	18.3	11.1
10	58.9	38.6	29.4	23.4	19.6	16.8	14.7	13.1	11.8	10.7	9.8	9.1	8.4	7.8	7.4	6.9	14.6	8.8
12	46.8	30.6	23.4	18.6	15.6	13.3	11.7	10.4	9.4	8.5	7.8	7.2	6.7	6.2	5.8	5.5	11.6	7.
14	37.	24.2	18.5	14.7	12.3	10.5	9.2	8.2	7.4	6.7	6.2	5.7	5.3	4.9	4.6	4.3	9.1	5.6
16	29.4	19.3	14.7	11.7	9.8	8.4	7.4	6.5	5.9	5.3	4.9	4.5	4.2	3.9	3.7	3.5	7.3	4.4
18	23.1	15.1	11.6	9.2	7.7	6.6	5.8	5.1	4.6	4.2	3.8	3.5	3.3	3.1	2.8	2.7	5.7	3.5
20	18.5	12.1	9.2	7.3	6.2	5.3	4.6	4.1	3.7	3.4	3.1	2.8	2.6	2.4	2.3	2.2	4.6	2.8

NUMBERS OF WIRES IN STRANDS.

Numbers, Brown & Sharpe gauge.

Circular mils	8	10	11	12	13	14	15	16	17	18	19	20	22	25	28	30	12 N. B. S. G.
1 000 000	61.	96.1	120.8	152.4	192.9	244.1	307.8	384.5	493.8	625.	771.6	976.6	1 562.	8 121.	6 299.	10 000.	92.5
950 000	58.	91.3	114.8	144.8	183.2	231.9	292.4	365.3	469.1	593.8	733.	927.8	1 484.	2 965.	5 984.	9 500.	87.9
900 000	54.9	86.5	108.7	137.2	173.6	219.7	277.	346.1	444.5	562.5	694.4	878.9	1 406.	2 809.	5 669.	9 000.	83.8
850 000	51.9	81.7	102.7	129.5	164.	207.5	261.6	326.8	419.8	531.3	655.9	830.1	1 328.	2 653.	5 354.	8 500.	76.
800 000	48.8	76.9	96.6	121.9	154.3	196.3	246.2	307.6	395.1	500.	617.3	781.3	1 250.	2 497.	5 039.	8 000.	74.
750 000	45.8	72.1	90.6	114.3	144.7	183.1	230.8	288.4	370.4	468.8	578.7	732.4	1 172.	2 241.	4 724.	7 500.	69.4
700 000	42.7	67.3	84.6	106.7	135.	170.9	215.5	269.2	345.7	437.5	540.1	683.6	1 094.	2 185.	4 409.	7 000.	64.8
650 000	39.7	62.5	78.5	99.1	125.4	158.7	200.1	249.9	321.	406.3	501.6	634.8	1 015.	2 029.	4 094.	6 500.	60.1
600 000	36.6	57.7	72.5	91.4	115.7	146.5	184.7	230.7	296.8	375.	463.	585.9	937.4	1 873.	3 779.	6 000.	55.5
550 000	33.6	52.9	66.4	83.8	106.1	134.3	169.3	211.5	271.6	343.8	424.4	537.1	859.3	1 717.	3 464.	5 500.	50.9
500 000	30.5	48.1	60.4	76.2	96.5	122.1	153.9	192.8	246.9	312.5	385.8	488.3	781.1	1 561.	3 149.	5 000.	46.3
450 000	27.5	43.2	54.4	68.6	86.8	109.8	138.5	173.	222.2	281.3	347.2	439.5	702.9	1 405.	2 835.	4 500.	41.6
400 000	24.4	38.4	48.3	61.	77.2	97.6	123.1	153.8	197.5	250.	308.6	390.6	624.9	1 248.	2 520.	4 000.	37.
350 000	21.4	33.6	42.3	53.3	67.5	85.4	107.7	134.6	172.8	218.8	270.1	341.8	546.8	1 092.	2 205.	3 500.	32.4
300 000	18.3	28.8	36.2	45.7	57.9	73.2	92.3	115.4	148.2	187.5	231.5	293.	468.7	936.3	1 890.	3 000.	27.8
250 000	15.3	24.	30.2	38.1	48.2	61.	77.	96.5	123.5	156.3	192.9	244.2	390.6	780.8	1 576.	2 500.	23.1

Number of wires in strands.

NUMBERS OF WIRES IN STRANDS.—(Continued.)

Numbers, Brown & Sharpe gauge.

Numbers, B. & S. G.	8	10	11	12	13	14	15	16	17	18	19	20	22	25	28	30
0.000	12.8	20.4	25.7	32.3	40.9	51.5	65.	81.9	103.3	130.3	164.3	207.1	329.4	660.4	1 324.	2 105.
000	10.2	16.2	20.4	25.7	32.3	40.9	51.5	65.	81.9	103.3	130.3	164.3	261.2	523.7	1 050.	1 670.
00	8.1	12.8	16.2	20.4	25.7	32.3	40.9	51.5	65.	81.9	103.3	130.3	207.1	415.3	882.7	1 324.
0	6.4	10.2	12.8	16.2	20.4	25.7	32.3	40.9	51.5	65.	81.9	103.3	164.3	329.4	660.4	1 050.
1	5.1	8.1	10.2	12.8	16.2	20.4	25.7	32.3	40.9	51.5	65.	81.9	130.3	261.2	523 7	832.7
2	4.	6.4	8.1	10.2	12.8	16.2	20.4	25.7	32.3	40.9	51.5	65.	103.3	207.1	415.3	660.4
3	3.2	5.1	6.4	8.1	10.2	12.8	16.2	20.4	25.7	32.3	40.9	51.5	81.9	164.3	329.4	523.7
4	2.5	4.	5.1	6.4	8.1	10.2	12.8	16.2	20.4	25.7	32.8	40.9	65.	130.3	261.2	415.3
5	2.	3.2	4.	5.1	6.4	8.1	10.2	12.8	16.2	20.4	25.7	32.3	51.5	103.3	207.1	329.4
6	1.6	2.5	3.2	4.	5.1	6.4	8.1	10.2	12.8	16.2	20.4	25.7	40.9	81.9	164.8	261.2
8	1.	1.6	2.	2.5	3.2	4.	5.1	6.4	8.1	10.2	12.8	16.2	25.7	51.5	108.3	164.3
10	....	1.	1.3	1.6	2.	2.5	3.2	4.	5.1	6.4	8.1	10.2	16.2	32.3	65.	108.3
12	....	....	....	1.	1.3	1.6	2.	2.5	3.2	4.	5.1	6.4	10.2	20.4	40.9	65.
14	....	....	....	....	....	1.	1.3	1.6	2.	2.5	3.2	4.	6.4	12.8	25.7	40.9
16	....	....	....	....	....	....	....	1.	1.3	1.6	2.	2.5	4.	8.1	16.2	25.7
18	....	....	....	....	....	....	....	....	....	1.	1.8	1.6	2.5	5.1	10.2	16.2
20	....	....	....	....	....	....	....	....	....	....	....	1.	1.6	3.2	6.4	10.2

Number of wires in strands.

IRON WIRE.

IN COMPARING tables of the weights of Galvanized Iron Wire it was found that the weights of the various sizes were not consistent with each other in the same table, and that no two tables seemed to agree in regard to the specific gravity of the material.

This table is calculated from the formula, weight per mile $= D^2 \times .013\,9$, which seems to be the most likely value for galvanized iron wire. This corresponds with a specific gravity of 7.73, and a weight per cubic foot of 483 pounds.

Steel wire is slightly heavier, and it is probable the constant in the above formula should be .014 for galvanized steel wire.

The following average values of the mile-ohm were used in calculating the resistance per mile at 68° F., the International ohm being the unit:

Kind of material.	Minimum.	Maximum.	Average.
E. B. B.,	4 500	4 800	4 700
B. B.,	5 300	6 000	5 600
Steel,	6 000	7 000	6 500

The breaking weight of any wire equals its weight per mile multiplied by 3 for E. B. B., 3.3 for B. B., or 3.7 for steel, all annealed and galvanized. This corresponds to 53 100 pounds, 58 410 pounds, and 65 490 pounds per square inch, respectively.

The strength of steel wire varies from 50 000 pounds per square inch to over 300 000 pounds, according to the kind of material and its treatment.

By taking 100 000 pounds per square inch as the breaking strain of steel wire, the breaking strain of any other wire may easily be computed from the table. For a wire of 80 000 pounds per square inch breaking strain, take eight-tenths of the tabulated breaking strain for that size wire at 100 000 pounds per square inch given in the table.

GALVANIZED IRON WIRE.

Numbers, B. W. G.	Diameters in mils.	Weights. Pounds.		Breaking weights. Pounds.		Resistance per mile in ohms.		
		1 000 feet.	One mile.	Iron.	Steel.	E. B. B.	B. B.	Steel.
0	340	304	1 607	4 821	9 079	2.93	3.42	4.05
1	300	237	1 251	3 753	7 068	3.76	4.4	5.2
2	284	212	1 121	3 368	6 335	4.19	4.91	5.8
3	259	177	932	2 796	5 268	5.04	5.9	6.97
4	238	149	787	2 361	4 449	5.97	6.99	8.26
5	220	127	673	2 019	3 801	6.99	8.18	9.66
6	203	109	573	1 719	3 237	8.21	9.6	11.35
7	180	85	450	1 350	2 545	10.44	12.21	14.43
8	165	72	378	1 134	2 138	12.42	14.53	17.18
9	148	58	305	915	1 720	15.44	18.06	21.35
10	134	47	250	750	1 410	18.83	22.04	26.04
11	120	38	200	600	1 131	23.48	27.48	32.47
12	109	31	165	495	933	28.46	33.3	39.36
13	95	24	125	375	709	37.47	43.85	51.82
14	83	18	96	288	541	49.08	57.44	67.88
15	72	13.7	72	216	407	65.23	76.33	90.21
16	65	11.1	59	177	332	80.03	93.66	110.7
17	58	8.9	47	141	264	100.5	120.4	139.
18	49	6.3	33	99	189	140.8	164.8	194.8

GALVANIZED IRON TELEGRAPH WIRE.

Western Union Telegraph company's specifications.
(Condensed).

"1. The wire to be soft and pliable, and capable of elongating 15 per cent. without breaking, after being galvanized.

"2. Great tensile strength is not required, but the wire must not break under a less strain than two and one-half times its weight in pounds per mile.

"3. Tests for ductility will be made as follows: The piece of wire will be gripped by two vises, 6 inches apart, and twisted. The full number of twists must be distinctly visible between the vises on the 6-inch piece. The number of twists in a piece of 6 inches in length not to be under 15.

"4. The weight per mile for the different gauge wires to be: for No. 4, 730 lbs.; No. 6, 540 lbs.; No. 8, 380 lbs.; No. 9, 320 lbs.; No. 10, 250 lbs., or, as near these figures as practicable.

"5. The electrical resistance of the wire in ohms per mile, at a temperature of 68° Fahrenheit, must not exceed the quotient arising from the dividing the constant number 4 800 by the weight of the wire in pounds per mile. The coëfficient .003 will be allowed for each degree Fahrenheit in reducing to standard temperature.

"6. The wire must be well galvanized, and capable of standing the following tests: The wire will be plunged into a saturated solution of sulphate of copper, and permitted to remain one minute, and then wiped clean. This process will be performed four times. If the wire appears black after the fourth immersion, it shows that the zinc has not been all removed, and that the galvanizing is well done; but if it has a copper color, the iron is exposed, showing that the zinc is too thin."

GALVANIZED IRON TELEGRAPH WIRE.
British Post-office specifications.

Diameters in mils.			Weights per mile. Pounds.			Strength and ductility.						Resistance per mile of standard size at 60° F. International ohms.
Required.	Maximum.	Minimum.	Required.	Maximum.	Minimum.	Minimum breaking weight. Pounds.	Twists in six inches.	Breaking weight, Not less than	Twists in six inches.	Breaking weight, Not less than	Twists in six inches.	
242	247	237	800	833	767	2 480	15	2 550	14	2 620	13	6.66
209	214	204	600	629	571	1 860	17	1 910	16	1 960	15	8.88
181	186	176	450	477	424	1 390	19	1 425	18	1 460	17	11.84
171	176	166	400	424	377	1 240	21	1 270	20	1 300	19	13.32
121	125	118	200	213	190	620	30	638	28	655	26	26.61

"The wire shall be well galvanized with zinc spelter, and this will be tested by an officer appointed by the Postmaster-General to inspect and test the wire, and hereinafter called the Inspecting Officer, taking samples from any piece or pieces and plunging them into a saturated solution of sulphate of copper, at 60° F., and allowing them to remain in the solution for one minute, when they are to be withdrawn and wiped clean. The galvanizing shall admit of this process being four times performed with each sample without there being, as there would be if the coating of zinc were too thin, any sign of a reddish deposit of metallic copper on the wire. Samples taken from pieces of the 800-lb. wire shall also bear bending around a bar 2¼ inches in diameter without any signs appearing of the zinc cracking or peeling off; the 600-lb. wire shall similarly bear bending around a bar 2¼ inches in diameter; the 450-lb. and 400-lb. wire around a bar 2 inches in diameter; and the 200-lb. wire around a bar 1½ inches in diameter."

The mile-ohm is 5 323.

GALVANIZED SUPPORTING STRANDS.

What weight per foot will a half-inch ordinary strand support if the strain is one-half the breaking weight, the span 120 feet, and the deflection .01 of the span or 1.2 feet?

One-half the breaking weight of a half-inch ordinary galvanized strand is 4 160 pounds. The value of S for above span and deflection, table page 50, is 1 500.2. Dividing 4 160 by 1 500.2 we find the total weight per foot to be 2.773 pounds. Deducting from this the weight per foot of the half-inch galvanized strand we have 2.263 pounds as the weight per foot of cable that this strand will support. While it is true that a factor of safety of two in this work is too small, yet the cables help in a great measure to carry their own weight. It is believed that galvanized strands will easily carry the loads indicated on page 39.

This strand is composed of seven wires, twisted together into a single strand.

Diameters in 32ds of an inch.	Weights per 100 feet. Pounds.	Estimated breaking strength. Pounds.	
		Ordinary.	Special.
16	51	8 320	16 640
15	48	7 500	15 000
14	37	6 000	12 000
12	30	4 700	9 400
10	21	3 300	6 600
9	18	2 600	5 200
8	11½	1 750	3 500
7	8¾	1 300	2 600
6	6½	1 000	2 000
5	4½	700	1 400
4	2¾	375	750
3	2	320	640

SUPPORTING CAPACITY OF GALVANIZED STRANDS.

Ordinary.

Diameters of strands in 32ds of an inch.	Spans in feet.								
	100	110	120	125	130	140	150	175	200
	Weights of 1 000 feet of cable. Pounds.								
16	2 818	2 516	2 263	2 152	2 050	1 867	1 709	1 391	1 154
15	2 520	2 247	2 020	1 920	1 827	1 663	1 520	1 234	1 130
14	2 030	1 812	1 630	1 550	1 476	1 344	1 230	1 001	900
12	1 580	1 409	1 266	1 204	1 146	1 043	953	774	640
10	1 110	899	890	846	805	733	670	544	450
9	860	765	680	652	620	563	513	414	340
8	585	521	468	445	423	385	352	285	235
7	433	385	346	329	313	284	260	210	172
6	337	300	270	257	245	223	204	165	137

Special.

Diameters of strands in 32ds of an inch.	Spans in feet.								
	100	110	120	125	130	140	150	175	200
	Weights of 1 000 feet of cable. Pounds.								
16	6 146	5 482	5 036	4 814	4 510	4 244	3 928	3 292	2 818
15	5 520	4 974	4 520	4 320	4 134	3 808	3 520	2 948	2 520
14	4 430	3 994	3 630	3 470	3 322	3 058	2 830	2 372	2 030
12	3 460	3 118	2 832	2 708	2 592	2 836	2 206	1 848	1 580
10	2 430	2 008	1 990	1 902	1 820	1 676	1 550	1 298	1 110
9	1 900	1 710	1 540	1 484	1 420	1 306	1 206	1 008	860
8	1 285	1 157	1 051	1 005	961	885	819	685	585
7	953	857	778	745	712	655	607	507	473
6	737	663	603	577	558	509	472	393	337

Dip = .01 of span.
Factor of safety of two.

CURRENTS.

FUSING EFFECTS OF CURRENTS.

Table giving the diameters of wires of various materials which will be fused by a current of given strength.

W. H. Preece, F.R.S.

$$d = \left(\frac{C}{a}\right)^{\frac{2}{3}}$$

Diameters in inches.

Current in amperes.	Copper, a=10 244.	Aluminum, a=7 585.	Platinum, a=5 172.	German silver, a=5 230.	Platinoid, a=4 750.	Iron, a=3 148.	Tin, a=1 642.	Tin-lead alloy, a=1 318.	Lead, a=1 379.
1	0.002 1	0.002 6	0.003 3	0.003 3	0.003 5	0.004 7	0.007 2	0.008 3	0.008 1
2	0.003 4	0.004 1	0.005 3	0.005 3	0.005 6	0.007 4	0.011 3	0.013 2	0.012 8
3	0.004 4	0.005 4	0.007	0.006 9	0.007 4	0.009 7	0.014 9	0.017 3	0.016 8
4	0.005 3	0.006 5	0.008 4	0.008 4	0.008 9	0.011 7	0.018 1	0.021	0.020 3
5	0.006 2	0.007 6	0.009 8	0.009 7	0.010 4	0.013 6	0.021	0.024 3	0.023 6
10	0.009 8	0.012	0.015 5	0.015 4	0.016 4	0.021 6	0.033 4	0.038 6	0.037 5
15	0.012 9	0.015 8	0.020 3	0.020 2	0.021 5	0.028 3	0.043 7	0.050 6	0.049 1
20	0.015 6	0.019 1	0.024 6	0.024 5	0.026 1	0.034 3	0.052 9	0.061 3	0.059 5
25	0.018 1	0.022 2	0.028 6	0.028 4	0.030 3	0.039 8	0.061 4	0.071 1	0.069
30	0.020 5	0.025	0.032 3	0.032	0.034 2	0.045	0.069 4	0.080 3	0.077 9
35	0.022 7	0.027 7	0.035 8	0.035 6	0.037 9	0.049 8	0.076 9	0.089	0.086 4
40	0.024 8	0.030 3	0.039 1	0.038 8	0.041 4	0.054 5	0.084	0.097 3	0.094 4
45	0.026 8	0.032 8	0.042 3	0.042	0.044 8	0.058 9	0.090 9	0.105 2	0.102 1
50	0.028 8	0.035 2	0.045 4	0.045	0.048	0.063 2	0.097 5	0.112 9	0.109 5
60	0.032 5	0.039 7	0.051 3	0.050 9	0.054 2	0.071 4	0.110 1	0.127 5	0.123 7
70	0.036	0.044	0.056 8	0.056 4	0.060 1	0.079 1	0.122	0.141 3	0.137 1
80	0.039 4	0.048 1	0.062 1	0.061 6	0.065 7	0.086 4	0.133 4	0.154 4	0.149 9
90	0.042 6	0.052	0.067 2	0.066 7	0.071 1	0.093 5	0.144 3	0.167 1	0.162 1
100	0.045 7	0.055 8	0.072	0.071 5	0.076 2	0.100 8	0.154 8	0.179 2	0.173 9
120	0.051 6	0.063	0.081 4	0.080 8	0.086 1	0.113 3	0.174 8	0.202 4	0.196 4
140	0.057 2	0.069 8	0.090 2	0.089 5	0.095 4	0.125 5	0.193 7	0.224 3	0.217 6
160	0.062 5	0.076 3	0.098 6	0.097 8	0.104 3	0.137 2	0.211 8	0.245 2	0.237 9
180	0.067 6	0.082 6	0.106 6	0.105 8	0.112 8	0.148 4	0.229 1	0.265 2	0.257 3
200	0.072 5	0.088 6	0.114 4	0.113 5	0.121	0.159 2	0.245 7	0.284 5	0.276
225	0.078 4	0.095 8	0.123 7	0.122 8	0.130 9	0.172 2	0.265 8	0.307 7	0.298 6
250	0.084 1	0.102 8	0.132 7	0.131 7	0.140 4	0.184 8	0.285 1	0 330 1	0.320 3
275	0.089 7	0.109 5	0.141 4	0.140 4	0.149 7	0.196 9	0.303 8	0.351 8	0.341 7
300	0.095	0.116 1	0.149 8	0.148 7	0.158 6	0.208 6	0.322	0.372 8	0.361 7

FUSING EFFECTS OF CURRENTS.—(Continued.)

Table showing the amperes required to fuse wires of various sizes and materials.

Number, N.B.S.G.	Diameter, d.	d^(3/2)	Copper, a=10.244	Aluminum, a=7.585	Platinum, a=5.172	German silver, a=5.230	Platinoid, a=4.750	Iron, a=3.148	Tin, a=1.642	Tin-lead alloy, a=1.318	Lead, a=1.379
14	0.08	0.022 627	231.8	171.6	117	118.3	107.5	71.22	37.15	29.82	31.2
16	0.064	0.016 191	165.8	122.8	83.73	84.68	76.9	50.96	26.58	21.34	22.32
18	0.048	0.010 516	107.7	79.75	54.37	54.99	49.96	33.1	17.27	13.86	14.5
20	0.036	0.006 831	69.97	51.18	35.33	35.72	32.44	21.5	11.22	9.002	9.419
22	0.028	0.004 685	48.	35.53	24.23	24.5	22.25	14.75	7.692	6.175	6.461
24	0.022	0.003 263	33.43	24.75	16.88	17.06	15.5	10.27	5.357	4.3	4.499
26	0.018	0.002 415	24.74	18.32	12.49	12.63	11.47	7.602	3.965	3.183	3.33
28	0.014 8	0.001 801	18.44	13.66	9.311	9.416	8.552	5.667	2.956	2.373	2.483
30	0.012 4	0.001 381	14.15	10.47	7.142	7.222	6.559	4.347	2.267	1.82	1.904
32	0.010 8	0.001 122	11.5	8.512	5.805	5.87	5.38	3.533	1.843	1.479	1.548

NOTE.—The size of "cut-outs," or fuses for electric-lighting circuits, can be taken at once from the first table. Pure copper wire makes the best and most reliable cut-out or fuse, and should never be less than one inch in length between the terminals to which it is fixed so as to prevent the cooling effect of the terminals.

HEATING EFFECTS OF CURRENTS.

A REPORT read before the Edison Convention, at Niagara Falls, August, 1889, by A. E. Kennelly, gives complete formulæ and tables based on experimental data, showing the heating effects of electric currents. This report was published in the *Electrical World*, beginning with the edition of November 23, 1889.

The tables in this book are taken from curves constructed from data given in the above report.

The table page 43 gives the rules of the various insurance companies, together with one column giving the current whose double would cause a rise of 40° C. This is the safe carrying capacity recommended in Kennelly's report.

The table page 44 gives the diameters of various wires and the current they will carry with a specified rise in temperature. The wires are insulated, and the conditions are similar to those met with in house wiring in mouldings or conduits.

The table page 45 is computed for bare wires suspended indoors, and gives the current carried with the corresponding rise in temperature.

The table page 46 is computed for outdoor wires, not insulated.

In these tables all wires are solid.

Insulation increases the current a wire will carry with a given rise in temperature, because the radiating surface is increased, and for the same reason a strand will carry a larger current than a solid wire.

One square inch of bright copper radiates .003 9 watts per degree Centigrade rise in temperature, and one square inch of blackened copper, .009 watts, under the same conditions. Convection seems to be dependent only on length, and may be taken at .053 watts per foot per degree Centigrade rise.

HEATING EFFECTS OF CURRENTS.

Insurance rules for carrying capacity of wires.

Numbers, B. & S. G.	Current, the double of which will cause a rise of 72° F.	National Electric Light association.	National Board of Fire Underwriters.		Associated Factory Mutual Insurance company.	Phoenix Fire Insurance company and Board of Trade rules of England.
			Concealed work.	Open work.		
0 000	174	175	218	312	175	.....
000	146	145	181	262	145	.....
00	123	120	150	220	120	105
0	103	100	125	185	100	83
1	88	95	105	156	85	66
2	73	70	88	131	70	52
3	61	60	75	110	60	41
4	52	50	63	92	50	33
5	43	45	53	77	45	26
6	36	35	45	65	35	21
7	31	30	.....	.....	30	16
8	26	25	33	46	25	13
10	18	20	25	32	20	8
12	13	15	17	23	15	5
14	9	10	12	16	10	3
16	6	5	6	8	5	2
18	5	.....	3	5	3	1

HEATING EFFECTS OF CURRENTS.—(Cont.)
Carrying capacity of insulated wires in mouldings.
(Kennelly's formula.)

Amperes.	Rise in temperature in degrees Centigrade.							
	5°	10°	15°	20°	30°	40°	50°	60°	70°
	Diameters of wires in mils.							
300	.....	.....	.....	.....	446	411	386	367	354
280	.....	.....	.....	.....	427	393	369	350	338
260	.....	.....	.....	450	409	375	352	333	321
240	.....	.....	.....	430	390	356	333	315	304
220	.....	.....	436	408	370	337	315	298	285
200	.....	448	414	386	350	317	295	280	268
190	.....	437	403	375	339	308	286	270	258
180	.....	425	391	364	328	298	277	260	249
170	.....	411	378	352	317	287	266	250	239
160	.....	398	364	340	305	276	256	241	229
150	445	383	351	326	293	265	244	230	218
140	431	370	338	312	281	253	232	220	206
130	417	354	322	300	269	240	220	208	195
120	400	339	308	285	255	228	208	195	182
110	383	322	292	270	240	214	195	182	170
100	362	302	276	253	223	200	182	168	158
90	343	284	259	237	208	185	168	154	143
80	322	264	240	218	192	169	153	139	130
70	300	242	220	198	174	152	139	123	116
60	275	220	195	175	155	135	122	108	101
50	250	195	175	152	132	118	104	91	86
40	217	169	144	128	110	95	85	75	70
30	178	136	115	100	85	73	66	58	54
20	132	100	71	69	59	50	45	40	37
10	78	58	42	35	30	.....	.....	.....	.....

HEATING EFFECTS OF CURRENTS.—(Cont.)
Bare copper in still air.

Amperes.	Rise in temperature, degrees Centigrade.							
	10°		20°		40°		80°	
	Bright.	Black.	Bright.	Black.	Bright.	Black.	Bright.	Black.
	Diameters of wires in mils.							
1 000	......	......	......	......	......	968	911	750
950	......	......	......	......	......	930	878	723
900	......	......	......	......	......	893	844	695
850	......	......	......	......	......	858	809	666
800	......	......	......	......	1 000	823	771	638
750	......	......	......	......	950	785	734	610
700	......	......	......	960	900	748	696	580
650	......	......	......	910	850	708	660	550
600	......	......	......	858	800	668	621	518
575	......	......	......	833	775	648	603	503
550	......	995	980	808	750	628	583	488
525	......	978	948	780	725	607	563	461
500	......	960	918	751	700	584	543	455
475	......	925	880	723	675	563	523	439
450	......	895	843	696	648	541	501	421
425	......	860	808	669	620	520	479	406
400	1 000	820	770	641	592	498	457	387
375	950	783	731	612	564	475	435	369
350	900	745	690	581	536	452	413	350
325	850	708	654	550	506	428	390	331
300	800	668	615	519	475	403	366	312
275	750	628	575	487	444	377	341	292
250	696	586	534	453	412	351	317	272
225	642	545	494	419	379	323	291	252
200	586	500	453	384	345	296	265	229
175	530	454	406	349	310	266	239	208
150	470	404	360	311	274	226	210	194
125	408	352	308	270	235	206	182	161
100	343	300	258	226	195	170	150	135
90	315	272	237	208	178	158	137	123
80	286	246	214	196	161	143	124	112
70	259	220	190	170	143	127	110	100
60	226	194	167	150	125	112	97	87
50	191	167	142	130	106	95	82	74
40	156	140	117	108	86	78	68	61
30	120	111	90	85	66	60	54	48
20	82	76	63	60	45	44	40	36
10	40	38	37	35	30	28	26	24

HEATING EFFECTS OF CURRENTS.—(Cont.)
Bare copper suspended outdoors.

Amperes	Rise in temperature, degrees Centigrade.							
	5°		10°		20°		40°	
	Bright.	Black.	Bright.	Black.	Bright.	Black.	Bright.	Black.
	Diameters of wires in mils.							
1 000	.....	.....	962	932	771	745	620	594
950	.....	.....	928	897	744	720	595	572
900	.....	.....	894	865	715	692	574	552
850	.....	.....	868	843	689	665	550	530
800	.....	.....	839	810	672	649	537	512
750	.....	975	804	775	643	620	515	495
700	963	933	767	739	613	591	491	472
650	916	889	729	703	582	561	467	449
600	869	837	690	665	554	532	442	426
575	845	813	671	647	538	517	429	414
550	820	789	650	627	522	501	417	402
525	795	764	630	609	506	487	404	389
500	770	740	610	589	489	470	390	376
475	745	719	589	569	478	455	377	363
450	719	693	568	548	453	438	368	350
425	690	667	546	526	436	422	349	336
400	661	638	524	504	418	406	334	322
375	632	610	502	484	399	377	319	309
350	601	581	478	462	380	360	304	295
325	571	552	453	439	362	342	289	279
300	540	522	428	415	342	326	273	264
275	509	492	404	392	321	309	257	249
250	477	460	378	367	300	290	240	222
225	445	430	351	343	280	270	223	215
200	410	399	324	316	259	250	205	198
175	373	365	296	289	235	227	186	180
150	334	329	267	258	211	202	166	161
125	295	290	235	226	185	177	145	144
100	254	248	202	193	157	152	123	120
90	236	230	186	178	145	140	114	111
80	216	212	171	164	132	128	104	102
70	198	192	155	150	120	116	94	91
60	177	170	137	132	107	104	83	80
50	155	147	119	115	92	87	72	70
40	130	124	100	96	77	73	62	59
30	104	100	78	75	61	58	50	45
20	73	70	54	53	43	40	34	30
10	40	38	27	26	20	18	16	14

SPANS.

THE formulæ used in calculating these tables of lengths and strains in spans of wire are those of a catenary of small deflection. They are given in Weisbach's Mechanics of Engineering, page 297, (seventh American edition, translated by Eckley B. Coxe, A. M.)

In these tables the horizontal strain at the center of the span is given. The strain at any other point equals the strain at the center plus the weight of a length of the wire equal to the perpendicular distance of that point from the lowest point of the wire in the span. For ordinary spans this is negligible. For any given wire the longest possible span is one where the deflection is about one-third of the span.

The effects of temperature on the strains of wires in spans is at first sight so great as to render the other considerations of little importance. The table, page 53, is calculated on the assumption that the supports of the spans are perfectly rigid under all conditions of strain and that the wire is inelastic. This is never true in practice. The changes in direction in a pole line afford a chance for the strains, due to a shortening of the wire by a fall in temperature, to be taken up by a bending of the supports.

If the elastic limit of hard-drawn copper wire of 60 000 pounds breaking strain be taken at 20 000 pounds, then S will equal 20 000 divided by 3.85, the weight of a piece of copper one foot long and one square inch in section. This makes S equal 5 195. Looking at the table of values of S, page 50, this value for a span of 130 feet comes between a deflection of .003 and .004. In the same way the allowable deflection for any other span of hard-drawn copper could be found or for any other material by substituting the proper terms for the elastic limit and the weight per foot given above.

The following gives the practice of some of the telegraph and telephone companies in their line construction:

SPECIFICATIONS FOR STANDARD CONSTRUCTION OF HARD-DRAWN COPPER.

Temperature in degrees Fahrenheit.	Spans in feet.					
	75	100	115	130	150	200
	Sag in inches.					
—30	1	2	2½	3⅜	4½	8
—10	1¼	2½	3	3¾	5	9
10	1½	2⅝	3½	4⅜	5¾	10¼
30	1¾	3	4	5⅛	6¾	12
60	2½	4¼	5½	7	9	15⅜
80	3⅛	5⅜	7	8⅝	11¼	18¾
100	4⅛	7	9	11	14	22¼

For spans between 400 and 600 feet, the dip shall be 1-40th of the span.

For spans between 600 and 1 000 feet, the dip shall be 1-30th of the span.

Another company uses 40 poles to the mile, and in the East allows three-inch dip at center of spans. In the West, where the variation of temperature is greater, 10 inches dip is allowed in summer, and 8 inches in the winter. This construction applies to both copper and iron wire, and has been found by actual experience to give satisfactory results.

The following formulæ were used in calculating the tables:

(1) $S \times W =$ horizontal strain on wire at center of span

(2) $$S = \frac{y^2}{2x} + \frac{x}{6}.$$

(3) $$l = y\left[1 + \tfrac{2}{3}\left(\frac{x}{y}\right)^2\right].$$

(4) $$x = 3S - \sqrt{9S^2 - 3y^2}.$$

(5) $$x = \sqrt{\frac{3yl - 3y^2}{2}}$$

In these formulæ

y = one-half span.
l = one-half length of wire in span.
x = deflection at center in same units as y.
w = weight per foot of wire.

Suppose we have a span of 200 feet of hard-drawn copper wire weighing one pound to 10 feet, and a deflection of two feet or .01 of the span.

(2) $\quad S = \left(\dfrac{100}{2}\right)^2 + \dfrac{2}{3}.$

$\quad\quad = 2\,500.33 \, +.$

(3) $\quad l = 100 \left[1 + \dfrac{3}{3}\left(\dfrac{2}{100}\right)^2 \right].$

$\quad\quad = 100.026\,6 \, +.$

$\quad 2\,l = 200.053 \, +.$

(4) $\quad x = 7501 - \sqrt{56\,265\,001 - 30\,000}.$

$\quad\quad = 2.$

(5) $\quad x = \sqrt{\dfrac{30\,008 - 30\,000}{2}}$

$\quad\quad = 2.$

In calculating the table, page 53, the deflection of the line was determined at $-10°$ F. by formula 4, the value of S being 30 000 divided by 3.85 or 7 792. For the other temperatures the length of the wire was calculated from the following formula:

$\quad$ Length $= l \left(1 + .000\,009\,3\,t\right).$

Here t is the difference in temperature in degrees Fahrenheit.

By formula 5 the deflection corresponding to the new length was found.

The coëfficients of linear expansion for each degree Fahrenheit are as follows:

$\quad$ Copper, .000 009 3.
$\quad$ Iron, .000 006 8.
$\quad$ Lead, .000 016.

STRAINS AT CENTERS OF SPANS RESULTING FROM A GIVEN DEFLECTION.

Deflections in decimal parts of spans.

Multipliers.

Spans in feet.	.001	.002	.003	.004	.005	.006	.007	.008	.009	.010	.015
10	1 250.001	625.008	416.671	312.506	250.006	208.343	178.588	156.263	138.908	125.016	83.358
20	2 500.008	1 250.006	833.343	625.013	500.016	416.686	357.166	312.526	277.807	250.033	166.716
30	3 750.005	1 875.01	1 250.015	937.52	750.025	625.03	535.749	468.79	416.711	375.05	250.075
40	5 000.006	2 500.013	1 666.686	1 250.026	1 000.033	833.373	714.382	625.053	555.615	500.066	333.433
50	6 250.008	3 125.016	2 083.358	1 562.533	1 250.041	1 041.716	892.915	781.316	694.619	625.083	416.791
60	7 500.01	3 750.02	2 500.03	1 875.04	1 500.05	1 250.06	1 071.498	937.58	833.423	750.1	500.15
70	8 750.011	4 375.023	2 916.701	2 187.546	1 750.058	1 458.403	1 250.081	1 093.843	972.327	875.116	583.508
80	10 000.013	5 000.026	3 333.373	2 500.053	2 000.066	1 666.746	1 428.664	1 250.106	1 111.231	1 000.133	666.866
90	11 250.015	5 625.03	3 750.045	2 812.56	2 250.075	1 875.09	1 607.247	1 406.37	1 250.135	1 125.15	750.225
100	12 500.016	6 250.033	4 166.716	3 125.066	2 500.083	2 083.433	1 785.83	1 562.633	1 389.088	1 250.166	833.583
110	13 750.018	6 875.036	4 583.398	3 437.573	2 750.091	2 291.776	1 964.414	1 718.896	1 527.942	1 375.183	916.941
120	15 000.02	7 500.04	5 000.06	3 750.08	3 000.1	2 500.12	2 142.997	1 875.16	1 666.846	1 500.2	1 000.3
130	16 250.021	8 125.043	5 416.731	4 062.586	3 250.108	2 708.463	2 321.58	2 031.423	1 806.75	1 625.216	1 083.658
140	17 500.023	8 750.046	5 833.408	4 375.093	3 500.116	2 916.806	2 500.163	2 187.686	1 944.654	1 750.233	1 167.016
150	18 750.025	9 375.05	6 250.075	4 687.6	3 750.125	3 125.15	2 678.746	2 343.95	2 083.558	1 875.25	1 250.376
160	20 000.026	10 000.053	6 666.746	5 000.106	4 000.133	3 333.493	2 857.329	2 500.213	2 222.462	2 000.266	1 333.733
170	21 250.028	10 625.056	7 083.418	5 312.613	4 250.141	3 541.836	3 085.912	2 656.476	2 361.366	2 125.283	1 417.091
180	22 500.03	11 250.06	7 500.09	5 625.12	4 500.15	3 750.18	3 214.495	2 812.74	2 500.269	2 250.3	1 500.45
190	23 750.031	11 875.063	7 916.761	5 937.626	4 750.158	3 958.523	3 393.078	2 969.003	2 639.173	2 375.316	1 583.808
200	25 000.033	12 500.066	8 333.433	6 250.133	5 000.166	4 166.866	3 571.661	3 125.266	2 778.077	2 500.333	1 667.166

STRAINS AT CENTERS OF SPANS RESULTING FROM A GIVEN DEFLECTION.—(Cont.)

Deflections in decimal parts of spans.

Spans in feet.	.020	.025	.030	.035	.040	.045	.050	.055	.060	.065	.070	.075
						Multipliers.						
10	62.533	50.041	41.716	35.772	31.316	27.852	25.083	22.818	20.933	19.339	17.973	16.791
20	125.066	100.083	83.433	71.545	62.633	55.705	50.166	45.637	41.866	38.678	35.947	33.583
30	187.6	150.125	125.15	107.317	93.90	83.538	75.25	68.456	62.8	58.017	53.921	50.375
40	250.133	200.166	166.866	143.09	125.266	111.411	100.333	91.275	83.733	77.356	71.896	67.166
50	312.666	250.208	208.583	178.863	156.583	139.263	125.416	114.094	104.666	96.695	89.869	83.958
60	375.2	300.25	250.3	214.635	187.900	167.116	150.5	136.913	125.6	116.084	107.042	100.75
70	437.733	350.291	292.016	250.408	219.216	194.969	175.583	159.732	146.533	135.373	125.816	117.541
80	500.266	400.333	333.733	286.18	250.533	222.822	200.666	182.551	167.466	154.712	143.79	134.333
90	562.8	450.375	375.45	321.953	281.850	250.674	225.75	206.37	188.4	174.051	161.764	151.125
100	625.333	500.416	417.166	357.726	313.166	278.527	250.833	228.189	209.333	193.391	179.738	167.916
110	687.866	550.458	458.883	393.498	344.483	306.38	275.916	251.008	230.266	212.73	197.711	184.708
120	750.4	600.5	500.6	429.271	375.800	334.233	301.	273.827	251.2	232.069	215.685	201.5
130	812.933	650.541	542.316	465.044	407.116	362.086	326.083	296.646	272.133	251.408	233.659	218.291
140	875.466	700.583	584.033	500.816	438.433	389.938	351.166	319.465	293.066	270.747	251.633	235.083
150	938.	750.625	625.75	536.589	469.750	417.791	376.25	342.284	314.	290.086	269.607	251.875
160	1 000.533	800.666	667.466	572.361	501.066	445.644	401.333	365.103	334.933	309.425	287.58	268.666
170	1 063.066	850.708	709.183	608.134	532.383	473.497	426.416	387.921	355.866	328.764	305.554	285.468
180	1 125.6	900.75	750.9	643.907	563.7	501.349	451.5	410.74	376.8	348.103	323.528	302.25
190	1 188.133	950.791	792.616	679.679	595.016	529.202	476.583	433.559	397.733	367.442	341.502	319.041
200	1 250.666	1 000.833	834.333	715.452	626.333	557.055	501.666	456.378	418.666	386.782	359.476	335.833

STRAINS AT CENTERS OF SPANS RESULTING FROM A GIVEN DEFLECTION.—(Cont.)

Deflections in decimal parts of spans.

Multipliers.

Spans in feet.	.080	.085	.090	.095	.100	.110	.120	.130	.140	.150	.160	.170	.180	.190	.200
10	15.758	14.847	14.038	13.316	12.666	11.546	10.616	9.832	9.161	8.583	8.079	7.636	7.244	6.895	6.583
20	31.516	29.695	28.077	26.632	25.333	23.093	21.233	19.664	18.323	17.166	16.158	15.272	14.488	13.791	13.166
30	47.275	44.542	42.116	39.948	38.	34.64	31.85	29.496	27.485	25.75	24.237	22.908	21.733	20.686	19.75
40	63.033	59.39	56.155	53.264	50.666	46.187	42.466	39.328	36.647	34.333	32.316	30.545	28.977	27.582	26.333
50	78.791	74.237	70.194	66.581	63.333	57.734	53.083	49.16	45.809	42.916	40.395	38.181	36.222	34.478	32.916
60	94.55	89.085	84.233	79.897	76.	69.281	63.7	58.992	54.971	51.5	48.476	45.817	43.466	41.373	39.5
70	110.308	103.932	98.272	93.213	88.666	80.828	74.316	68.824	64.133	60.083	56.554	53.453	50.711	48.269	46.083
80	126.066	118.78	112.311	106.529	101.333	92.375	84.933	78.656	73.295	68.666	64.633	61.09	57.955	55.164	52.666
90	141.825	133.627	126.35	119.846	114.	103.922	95.55	88.488	82.457	77.25	72.712	68.726	65.199	62.06	59.25
100	157.583	148.475	140.388	133.162	126.666	115.469	106.166	98.32	91.619	85.833	80.791	76.362	72.444	68.956	65.833
110	173.341	163.323	154.427	146.478	139.333	127.016	116.783	108.152	100.78	94.416	88.87	83.999	79.688	75.851	72.416
120	189.1	178.17	168.466	159.794	152.	138.563	127.4	117.984	109.942	103.	96.95	91.635	86.983	82.747	79.
130	204.858	193.018	182.505	173.11	164.666	150.11	138.016	127.816	119.104	111.583	105.029	99.271	94.177	89.642	85.583
140	220.616	207.865	196.544	186.427	177.333	161.657	148.633	137.648	128.266	120.166	113.108	106.907	101.422	96.538	92.166
150	236.375	222.713	210.583	199.743	190.	173.204	159.25	147.48	137.428	128.75	121.187	114.544	108.666	103.434	98.75
160	252.133	237.56	224.622	213.059	202.666	184.751	169.866	157.312	146.59	137.333	129.266	122.18	115.911	110.329	105.333
170	267.891	252.408	238.661	226.375	215.333	196.298	180.483	167.144	155.752	145.916	137.345	129.816	123.155	117.225	111.916
180	283.65	267.255	252.7	239.692	228.	207.845	191.1	176.976	164.914	154.5	145.425	137.452	130.399	124.121	118.5
190	299.408	282.103	266.738	253.008	240.666	219.392	201.716	186.808	174.076	163.083	153.604	145.089	137.644	131.016	125.083
200	315.166	296.95	290.777	266.324	253.333	230.989	212.338	196.641	183.238	171.666	161.583	152.725	144.883	137.912	131.666

RULE.—To find strain in pounds on wire of given span and deflection, multiply numbers in column answering to span and deflection by the weight per foot of wire.

TEMPERATURE EFFECTS IN SPANS.

Spans in feet.	Temperature in degrees Fahrenheit.								
	-10°	30°	40°	50°	60°	70°	80°	90°	100°
	Deflections in inches.								
50	.5	6	8	9	9	10	11	11	12
60	.7	8	10	11	11	12	13	13	14
70	1.	10	11	12	13	14	15	15	17
80	1.2	11	13	14	15	16	17	18	19
90	1.6	13	14	16	17	18	19	20	21
100	1.9	14	16	17	19	20	21	23	24
110	2.3	16	18	19	21	22	24	25	26
120	2.8	17	19	21	22	24	26	27	28
130	3.2	19	21	23	25	26	28	29	31
140	3.7	20	23	25	27	28	30	32	33
150	4.3	22	24	26	28	30	32	34	36
160	4.9	23	26	28	30	32	34	36	38
170	5.5	25	28	30	32	35	37	38	40
180	6.2	26	29	32	34	37	39	41	43
190	7.	28	31	34	36	39	41	43	45
200	7.7	31	33	36	38	41	43	45	48

Hard-drawn copper wire, 60 000 pounds strength per square inch.

Strain at −10° F., 30 000 pounds per square inch.

TOTAL LENGTHS OF WIRES IN SPANS.

Deflections in decimal parts of spans.

Lengths of wires.

Spans in feet.	.010	.015	.020	.025	.030	.035	.040	.045	.050	.065	.060	.070	.080
10	10.002	10.006	10.01	10.016	10.024	10.032	10.042	10.064	10.066	10.08	10.096	10.13	10.17
20	20.005	20.012	20.021	20.083	20.048	20.065	20.065	20.108	20.133	20.161	20.192	20.261	20.341
30	30.008	30.018	30.032	30.05	30.072	30.098	30.128	30.162	30.2	30.242	30.288	30.392	30.512
40	40.01	40.024	40.042	40.066	40.096	40.13	40.17	40.216	40.266	40.322	40.384	40.522	40.682
50	50.013	50.03	50.053	50.083	50.12	50.163	50.213	50.27	50.333	50.403	50.48	50.653	50.853
60	60.016	60.036	60.064	60.1	60.144	60.196	60.256	60.324	60.4	60.484	60.576	60.784	61.024
70	70.018	70.042	70.074	70.116	70.168	70.228	70.298	70.378	70.466	70.564	70.672	70.914	71.194
80	80.021	80.048	80.085	80.133	80.192	80.261	80.341	80.432	80.533	80.645	80.768	81.045	81.365
90	90.024	90.054	90.096	90.15	90.216	90.294	90.384	90.486	90.6	90.726	90.864	91.176	91.536
100	100.026	100.06	100.106	100.166	100.24	100.326	100.426	100.54	100.666	100.806	100.96	101.306	101.706
110	110.029	110.066	110.117	110.183	110.264	110.359	110.469	110.594	110.733	110.887	111.056	111.437	111.877
120	120.032	120.072	120.128	120.2	120.288	120.392	120.512	120.648	120.8	120.968	121.152	121.568	122.048
130	130.034	130.078	130.149	130.216	130.312	130.424	130.564	130.702	130.866	131.048	131.248	131.698	132.218
140	140.037	140.084	140.149	140.233	140.336	140.457	140.597	140.756	140.933	141.129	141.344	141.829	142.389
150	150.04	150.09	150.16	150.25	150.36	150.49	150.64	150.81	151.	151.21	151.44	151.96	152.56
160	160.04	160.096	160.17	160.266	160.384	160.522	160.682	160.864	161.066	161.29	161.536	162.09	162.73
170	170.045	170.102	170.181	170.283	170.408	170.555	170.725	170.918	171.133	171.371	171.632	172.221	172.901
180	180.048	180.108	180.192	180.3	180.432	180.588	180.768	180.972	181.2	181.452	181.728	182.352	183.072
190	190.05	190.114	190.202	190.316	190.456	190.62	190.81	191.026	191.266	191.532	191.824	192.482	193.242
200	200.053	200.12	200.213	200.333	200.48	200.653	200.853	201.08	201.333	201.613	201.92	202.613	203.418

TOTAL LENGTHS OF WIRES IN SPANS.—(*Continued.*)

Deflections in decimal parts of spans.

Lengths of wires.

Spans in feet.	.090	.100	.110	.120	.130	.140	.150	.160	.170	.180	.190	.200
10	10.216	10.266	10.322	10.384	10.45	10.522	10.6	10.682	10.77	10.864	10.962	11.066
20	20.432	20.533	20.645	20.768	20.901	21.045	21.2	21.365	21.541	21.728	21.925	22.133
30	30.648	30.8	30.968	31.152	31.352	31.568	31.8	32.048	32.312	32.592	32.888	33.2
40	40.864	41.066	41.29	41.536	41.802	42.09	42.4	42.73	43.082	43.456	43.85	44.266
50	51.08	51.333	51.613	51.92	52.258	52.613	53.	53.413	53.853	54.32	54.818	55.333
60	61.296	61.6	61.936	62.304	62.704	63.136	63.6	64.096	64.624	65.184	65.776	66.4
70	71.512	71.866	72.258	72.688	73.154	73.658	74.2	74.778	75.394	76.048	76.738	77.466
80	81.728	82.133	82.581	83.072	83.605	84.181	84.8	85.461	86.165	86.912	87.701	88.533
90	91.944	92.4	92.904	93.456	94.056	94.704	95.4	96.144	96.936	97.776	98.664	99.6
100	102.16	102.666	103.226	103.84	104.506	105.226	106.	106.826	107.706	108.64	109.626	110.666
110	112.376	112.933	113.549	114.224	114.957	115.749	116.6	117.509	118.477	119.504	120.589	121.733
120	122.592	123.2	123.872	124.608	125.408	126.272	127.2	128.192	129.248	130.368	131.552	132.8
130	132.808	133.466	134.194	134.992	135.858	136.794	137.8	138.874	140.018	141.232	142.514	143.866
140	143.024	143.733	144.517	145.376	146.309	147.317	148.4	149.557	150.789	152.096	153.477	154.983
150	153.24	154.	154.84	155.76	156.76	157.84	159.	160.24	161.56	162.96	164.44	166.
160	163.456	164.266	165.162	166.144	167.21	168.362	169.6	170.922	172.33	173.824	175.402	177.066
170	173.672	174.533	175.485	176.528	177.661	178.886	180.2	181.605	183.101	184.688	186.365	188.133
180	183.888	184.8	185.808	186.912	188.112	189.408	190.8	192.288	193.872	195.552	197.328	199.2
190	194.104	195.066	196.13	197.296	198.562	199.93	201.4	202.97	204.642	206.416	208.29	210.166
200	204.32	205.333	206.453	207.68	209.013	210.453	212.	213.653	215.413	217.28	219.253	221.333

WEATHERPROOF WIRE.

Our Weatherproof wire is put on reels in long lengths, and has a hard, smooth finish, presenting the least possible chance for adherence of ice and snow. We keep in stock all sizes given in the accompanying table, to 0 000 B. & S., in both double and triple braid.

In the Stranded wires, we keep only the most commonly used sizes. We make this Feed Wire Strand either concentric or cable-laid, as desired.

FIRE AND WEATHERPROOF WIRE.

For interior work, we manufacture a Fire and Weatherproof insulation. Full information concerning weights, diameters and prices furnished on application.

UNDERWRITERS' WIRE.

Underwriters' wire seems to be used chiefly for inside work. Its weight is about the same as double-braid Weatherproof.

WEATHERPROOF IRON WIRE.

We keep in stock 10, 12 and 14 B. W. G., both double and triple braid.

Numbers, B. W. G.	Weights per mile. Pounds.		Lengths in coils. Miles.
	Double braid.	Triple braid.	
4	997	1 102	$\frac{1}{8}$
6	713	773	$\frac{1}{8}$
8	483	548	$\frac{1}{4}$
9	403	464	$\frac{1}{8}$
10	350	410	$\frac{1}{8}$
12	240	265	$\frac{1}{2}$
14	150	176	$\frac{1}{2}$

WEATHERPROOF WIRE.

Numbers, B. & S. G.	Outside diameters in 32ds inch	Double braid. Weights. Pounds.		Outside diameters in 32ds inch	Triple braid. Weights. Pounds.		Approximate weights. Pounds.	
		1 000 feet.	Mile.		1 000 feet.	Mile.	Reel.	Coil.
0 000	20	716	3 781	24	775	4 092	2 000	250
000	18	575	8 086	22	630	3 326	2 000	250
00	17	465	2 455	18	490	2 587	500	250
0	16	375	1 980	17	400	2 112	500	250
1	15	285	1 505	16	306	1 616	500	250
2	14	245	1 294	15	268	1 415	500	250
3	13	190	1 008	14	210	1 109	500	250
4	11	152	808	12	164	866	250	125
5	10	120	634	11	145	766	260	130
6	9	98	518	10	112	591	275	140
8	8	66	349	9	78	412	200	100
10	7	45	238	8	55	290	200	100
12	6	30	158	7	35	185	.....	25
14	5	20	106	6	26	137	.....	25
16	4	14	74	5	20	106	.....	25
18	3	10	53	4	16	85	.....	25

STRANDED WEATHERPROOF FEED WIRE.

Circular mils.	Outside diameters. Inches.	Weights. Pounds.		Approximate length on reels. Feet.
		1 000 feet.	Mile.	
1 000 000	1½	3 550	18 744	800
900 000	1 7/16	3 215	16 975	800
800 000	1 3/8	2 880	15 206	850
750 000	1 5/16	2 713	14 325	850
700 000	1 9/32	2 545	13 488	900
650 000	1¼	2 378	12 556	900
600 000	1 3/16	2 210	11 668	1 000
550 000	1 5/32	2 043	10 787	1 200
500 000	1 1/8	1 875	9 900	1 320
450 000	1 3/32	1 703	8 992	1 400
400 000	1 1/16	1 580	8 078	1 450
350 000	1	1 358	7 170	1 500
300 000	15/16	1 185	6 257	1 600
250 000	27/32	1 012	5 343	1 600

The table is calculated for concentric strands. Rope-laid strands are larger.

RUBBER WIRE.

WE MANUFACTURE rubber insulated wires for all purposes, including wires and cables for aerial, underground, and submarine use. The copper conductor is tinned, and then covered with a cement of pure rubber, which causes the succeeding coat of rubber to adhere firmly to the wire. This layer consists of white rubber without sulphur. Over this is a layer of vulcanized rubber, and the whole is covered with a finishing braid of cotton saturated with a Weatherproof compound, which protects the rubber from mechanical injury, and from the action of the air. A poor quality of rubber insulation is inferior to Weatherproof, and we would recommend our Fire and Weatherproof insulation for inside work, rather than an inferior rubber wire.

A good rubber wire should have its conductor central, the insulation should adhere firmly to the wire, it should not crack or become brittle after use, and it should show, after immersion in water for twenty-four hours, the same insulation resistance per mile as when tested after being first put in water. The absolute number of megohms per mile depends on the age of the rubber used, together with other details of manufacture, and is not always a sure guide to the quality of the insulation. Uniformity of insulation among several coils of wire made at the same time, or among the various conductors of a cable, is a much more valuable aid in detecting a poor piece of wire, as in this case an insulation lower than the average shows a local defect, which, in time, will be likely to cause trouble.

CRESCENT RUBBER WIRE
Stranded conductors.

Numbers, B. & S. G.	Circular mils.	Outside diameters. Inches.	Weights per 1 000 feet. Pounds.	Sizes of wires in strands. B. & S. G.	
				Regular.	Flexible.
....	1 000 000	1 7/8	3 690	8	12
....	900 000	1 11/16	3 370	8	12
....	800 000	1 5/8	3 020	8	12
....	700 000	1 7/16	2 685	10	12
....	600 000	1 5/16	2 345	10	12
....	500 000	1 3/16	1 885	10	14
....	450 000	1 1/8	1 723	10	14
....	400 000	1	1 560	10	14
....	350 000	15/16	1 378	10	14
....	300 000	7/8	1 155	10	14
	250 000	27/32	995	10	14
0 000	....	25/32	866	10	15
000	....	23/32	725	10	15
00	....	21/32	613	11	15

Numbers, B. & S. G.	Outside diameters in 32ds of an inch.		Weights per 1 000 feet. Pounds.	Sizes of wires in strand. B. & S. G.	
	Solid.	Stranded.		Regular.	Flexible.
0	18	20	489	12	16
1	16	18	398	12	16
2	14	15	309	12	18
3	13	14	244	13	18
4	12	13	198	14	20
5	11	12	168	15	20
6	10	11	146	16	20
8	9	10	106	18	22
10	8	8	77	20	25
12	7	7	55	20	25
14	6	6	35	21	25
16	5	5	25	23	25
18	4	4	20	25	25

MAGNET WIRE.

THE BARE COPPER intended for Magnet wire is specially drawn and annealed, great care being taken to have it true to gauge, and soft.

A difference from the standard, of one mil, is allowed on sizes larger than No. 10 B. & S. G.; from No. 10 to No. 14, three-fourths of a mil variation is allowed, and any wire smaller than No. 14, one-half a mil variation is allowed.

The insulation is smooth and uniform, and is kept true to gauge to within one mil of the required diameter.

We manufacture any special kind of Magnet wire required, flats, squares and strands.

We understand that a No. 6 B. & S. square Magnet wire measures 162 x 162 mils.

Flats are designated by their width and thickness. Thus a flat Magnet wire 340 mils wide and 40 mils thick would be designated as a 340 x 40 flat Magnet wire.

Strands can be furnished of any size, insulated with double or triple windings of cotton, or any combination of braids and windings that may be desired.

MAGNET WIRE.

Numbers, B. & S. G.	Diameter drawn. Mils.	Outside diameters. Mils.		Approximate weights on reels. Pounds.
		Double.	Single.	
0	325	343	337	200
1	289	307	301	200
2	258	276	270	200
3	229	247	241	200
4	204	222	216	200
5	182	200	194	200
6	162	178	172	200
7	144	160	154	200
8	128	142	137	200
9	114	126	122	200
10	102	112	108	200
11	91	101	97	200
12	81	91	87	200
13	72	81	78	160
14	64	73	70	160
15	57	66	63	50
16	51	60	57	50
17	45	54	51	50
18	40	49	46	50
19	36	45	42	50

GERMAN SILVER WIRE.

Numbers, B. & S. G.	Resistance per 1 000 feet.		Numbers, B. & S. G.	Resistance per 1 000 feet.	
	18 per centum.	30 per centum.		18 per centum.	30 per centum.
6	7.20	11.21	22	295.38	459.48
7	9.12	14.18	23	370.26	575.96
8	11.54	17.95	24	468.18	728.28
9	14.55	22.63	25	590.22	918.12
10	18.18	28.28	26	748.08	1 163.68
11	22.84	35.53	27	937.98	1 459.08
12	28.81	44.82	28	1 191.24	1 853.04
13	36.48	56.75	29	1 481.22	2 304.12
14	46.17	71.82	30	1 891.8	2 942.8
15	58.21	90.55	31	2 388.6	3 715.6
16	72.72	113.12	32	2 955.6	4 597.6
17	93.40	145.29	33	3 751.2	5 835.2
18	118.20	183.87	34	4 764.6	7 411.6
19	145.94	227.02	35	6 031.8	9 382.8
20	184.68	287.28	36	7 565.4	11 768.4
21	232.92	362.32	...	....	....

The resistance of German silver wire varies according to the method of manufacture and the materials used.

From actual tests on wire with eighteen per centum of nickel, extending over ten years, it seems that eighteen times the resistance of copper, at 75° F., represents very closely the resistance of this alloy. This value is rather under than over the average results of the tests.

For the thirty per centum alloy, we have to depend on the results of a single series of tests, and while the results are believed to be correct, they are not as reliable as those given for the eighteen per centum German silver wire. We take the resistance of the thirty per centum alloy at twenty-eight times the resistance of copper, at 75°.

The International ohm is taken as the unit of resistance.

OFFICE WIRES.

Office wire is usually made with a wind and a braid of cotton saturated with paraffine. It is sometimes required with a double braid or triple braid of cotton. The most common colors are red and white. Any combination of colors can be furnished.

Damp-proof Office wire has the inside wind saturated with black Weatherproof compound, while the outside finish is the same as ordinary Office wire.

Annunciator wire has a covering consisting of two wraps of cotton saturated with paraffine. The outer covering is made in solid colors or combination of two colors.

Double conductors for house wiring are of various kinds.

Two conductors twisted together, without any outside cover, form a convenient method of wiring for bells, telephones, etc. These conductors may be 18 B. & S., with double braid Weatherproof or with Annunciator insulation.

Two-conductor Office wire may be two Office wires laid side by side and covered with a two-colored Office braid, or it may consist of two Annunciator wires so insulated.

Weatherproof cables consist of 18 B. & S. G. Annunciator wires, twisted into a cable and covered with rubber tape and a braid of cotton saturated with Weatherproof insulation. They weigh about ten pounds per 1 000 feet per conductor. For work inside building, in dry places, the rubber tape may be omitted, and the finishing braid made any color to correspond with the woodwork.

Lamp cord is furnished in silk or cotton insulation. Green and yellow is the standard color combination.

Numbers, B. & S. G.	Weights per 1 000 feet.		Sizes of Lamp cord.	
	Office wire.	Annunciator wire.	Silk.	Cotton.
14	17	15	$\frac{1}{4}$	$\frac{1}{16}$
16	12	10	$\frac{1}{16}$	$\frac{1}{4}$
18	9	7	$\frac{1}{16}$	$\frac{1}{16}$
20	7	$4\frac{1}{2}$	$\frac{1}{8}$	$\frac{1}{16}$

POWER CABLES.

WE MANUFACTURE power and electric-light cables, with jute, paper or rubber insulation. The thickness and kind of insulation depend on the use for which the cable is intended. The table of diameters and weights is based on $\frac{3}{16}$ insulation on a side, and is approximately correct for any kind of insulation.

Specifications for Underground Cable of 500 000 C. M.

1. Copper Conductor.

The conductor shall consist of 47 wires, each 104 mils in diameter, and shall weigh not less than 1.525 pounds per foot. The copper used shall have a conductivity of not less than 98 per cent.

2. Insulation.

The insulation shall consist of paper not less than $\frac{3}{16}$ thick, and shall form a wall of uniform thickness around the conductor.

3. Sheath.

The insulated conductor shall be enclosed in a pipe composed of lead and tin. The amount of tin shall not be less than 2.9 per cent. The pipe shall be formed around the core, and shall be free from holes or other defects, and of uniform thickness and composition.

4. Insulation Resistance.

The insulation resistance shall be not less than 300 megohms per mile, at 60° F.

POWER CABLES.

Numbers, B. & S. G.	Circular mils.	Outside diameters. Inches.	Weights, 1 000 feet. Pounds.
....	1 000 000	1 13/16	6 685
....	900 000	1 11/16	6 228
....	800 000	1 11/16	5 773
....	750 000	1 5/8	5 543
....	700 000	1 9/16	5 815
....	650 000	1 9/16	5 088
....	600 000	1 17/32	4 857
....	550 000	1 1/2	4 630
....	500 000	1 7/16	4 278
....	450 000	1 3/8	3 923
....	400 000	1 11/32	3 619
....	350 000	1 5/16	3 416
....	300 000	1 1/4	3 060
....	250 000	1 7/32	2 732
0 000	211 600	1 3/16	2 533
000	168 100	1 1/16	2 300
00	133 225	1	2 021
0	105 625	15/16	1 772
1	83 521	29/32	1 633
2	66 564	7/8	1 482
3	52 441	13/16	1 360
4	41 616	3/4	1 251
6	26 244	11/16	1 046

TELEPHONE CABLES.

Lead-encased for underground or aerial use.

THE INSULATION of these cables is dry paper. We manufacture several styles of 19 B. & S. G., 20 B. & S. G., and 22 B. & S. G., according to the use for which they are intended. The most common size is 19 B. & S. G. We also supply terminals and hangers. To determine the size supporting strand to use with these cables, consult tables page 39.

Specifications for Telephone Cables.

1. CONDUCTORS.

Each conductor shall be .035 89 inches in diameter, (19 B. & S. G.,) and have a conductivity of 98 per cent. of that of pure soft copper.

2. CORE.

The conductor shall be insulated, twisted in pairs, the length of the twist not to exceed three inches, and formed into a core arranged in reverse layers.

3. SHEATH.

The core shall be enclosed in a pipe composed of lead and tin, the amount of the tin shall be not less than $2\frac{1}{2}$ per cent. The pipe shall be formed around the core, and shall be free from holes or other defects, and of uniform thickness and composition.

4. ELECTROSTATIC CAPACITY.

The average electrostatic capacity shall not exceed .080 of a microfarad per mile, each wire being measured against all the rest and a sheath grounded; the electrostatic capacity of any wires so measured shall not exceed .085 of a microfarad per mile.

5. INSULATION RESISTANCE.

Each wire shall show an insulation of not less than 500 megohms per mile, at 60° F., when laid, spliced and connected to terminal ready for use; each wire being measured against all the rest and sheath grounded.

6. CONDUCTOR RESISTANCE.

Each conductor shall have a resistance of not more than 47 B. A. ohms, at 60° F., for each mile of cable, after the cable is laid and connected to the terminals.

TELEPHONE CABLES.

Number pairs.	Outside diameters. Inches.	Weights, 1 000 feet. Pounds.
1	5/16	214
2	3/8	302
3	1/2	515
4	9/16	629
5	5/8	747
6	3/4	877
7	7/8	912
10	13/16	1 214
12	15/16	1 375
15	1	1 566
18	1 1/16	1 758
20	1 1/8	1 940
25	1 3/16	2 332
30	1 7/8	2 748
35	1 1/2	2 985
40	1 5/16	3 176
45	1 5/8	3 365
50	1 3/4	3 678
55	1 13/16	3 867
60	1 7/8	4 055
65	1 15/16	4 241
70	2	4 430
80	2 1/8	4 804
90	2 1/4	5 180
100	2 3/8	5 505

TELEGRAPH CABLES.
Lead-encased for underground use.

THESE cables are made of either rubber, cotton or paper insulation. The sizes and weights are approximately correct for rubber and cotton insulation. Both sizes and weights are slightly reduced for paper insulation. In all cases the cables are lead-encased.

Specifications for Telegraph Cables.

1. CONDUCTORS.

Each conductor shall be .064 inches in diameter, (14 B. & S. G.,) and have a conductivity of 98 per cent. of that of pure copper.

2. CORE.

The conductors shall be insulated to $\frac{1}{32}$ with cotton, and formed into a core arranged in reverse layers. This core shall be dried and saturated with approved insulating compound.

3. SHEATH.

The core shall be enclosed in a pipe composed of lead and tin. The amount of tin shall not be less than 2.9 per cent. The pipe shall be formed around the core, and shall be free from holes or other defects, and of uniform thickness and composition.

4. INSULATION RESISTANCE.

The wire shall show an insulation of not less than 300 meg-ohms per mile, at 60° F., when laid, spliced and connected to terminals ready for use, each wire being measured against all the rest and the sheath grounded.

5. CONDUCTOR RESISTANCE.

Each conductor shall have a resistance of not more than 28 International ohms, at 60° F., for each mile of cable, after the cable is laid and connected up to the terminals.

TELEGRAPH CABLES.

Number conductors.	14 B. & S. G. Insulated to 1/32.		16 B. & S. G. Insulated to 1/32.		18 B. & S. G. Insulated to 1/32.	
	Outside diameters. Inches.	Weights, 1 000 feet.	Outside diameters. Inches.	Weights, 1 000 feet.	Outside diameters. Inches.	Weights, 1 000 feet.
1	3/8	306	3/8	299	3/8	291
2	7/16	438	7/16	421	13/32	356
3	1/2	573	1/2	546	7/16	421
4	5/8	810	17/32	670	15/32	486
5	3/4	972	5/8	793	1/2	551
6	13/16	1 132	11/16	946	17/32	616
7	7/8	1 295	3/4	965	9/16	681
10	15/16	1 512	13/16	1 155	5/8	820
12	1 1/16	1 873	7/8	1 327	3/4	978
15	1 3/16	2 263	15/16	1 518	13/16	1 148
18	1 1/4	2 523	1 1/16	1 880	7/8	1 318
20	1 5/16	2 756	1 1/8	2 076	15/16	1 477
25	1 7/16	3 250	1 3/16	2 496	1	1 690
30	1 9/16	3 515	1 5/16	2 768	1 1/16	1 908
35	1 11/16	3 910	1 7/16	3 040	1 5/32	2 116
40	1 3/4	4 175	1 1/2	3 312	1 1/4	2 330
45	1 13/16	4 441	1 17/32	3 533	1 9/32	2 471
50	1 15/16	4 835	1 5/8	3 755	1 5/16	2 628
55	2	5 100	1 11/16	3 978	1 3/8	2 866
60	2 1/16	5 365	1 3/4	4 200	1 7/16	3 104
65	2 1/8	5 631	1 13/16	4 422	1 15/32	3 245
70	2 3/16	5 897	1 7/8	4 644	1 1/2	3 402
80	2 5/16	6 408	2	5 087	1 5/8	3 798
90	2 7/16	6 916	2 1/16	5 402	1 11/16	4 027
100	2 9/16	7 375	2 1/8	5 720	1 3/4	4 275

AERIAL CABLES.

THESE cables are made from double-coated rubber wire, taped. After standing, the cable is double-taped and covered with tarred jute, over which is placed a braid of heavy cotton saturated with Weatherproof compound. This outside covering protects the rubber from the action of the air and from mechanical injury. The separate wires are tested in water, and no wire is used which will not fully meet a water test. The result is a cable which will work under water as well as on a pole line, if there is no danger of mechanical injury. The ordinary size for telegraphic work is 14 B. & S., insulated to $\frac{8}{32}$. A trace wire can be placed in each layer, if desired.

The size galvanized strand to support these cables may be found from the table page 39. Suppose the span is 130 feet and a 10-conductor 14 B. & S. G. Aerial cable is used, then from these tables it is seen a $\frac{1}{4}$-inch ordinary galvanized strand will support a cable weighing 423 pounds per 1 000 feet, with a 130-foot span.

Specifications for 14 B. & S. Aerial Cable.

1. CONDUCTORS.

Each conductor shall be .064 inches in diameter, (14 B. & S. G.,) and have a conductivity of 98 per cent. of that of pure copper.

2. CORE.

The conductors shall be insulated to $\frac{3}{32}$ with rubber and tape, and formed into a core arranged in reverse layers.

3. PROTECTIVE COVERING.

The core shall be covered with two wraps of friction tape and one wrap of tarred jute. Over this there shall be a braid saturated with Weatherproof compound.

4. INSULATION RESISTANCE.

Each wire shall show an insulation resistance of not less than 300 megohms per mile, at 60° F., after being immersed in water 24 hours. This test shall be made on the core after all the conductors are laid up, but before the outside coverings are put on.

5. CONDUCTOR RESISTANCE.

Each conductor shall have a resistance of not more than 28 International ohms, at 60° F., for each mile of cable.

AERIAL CABLES.
Rubber insulation.

Number conductors.	14 B. & S. G. Insulated to 1/32.		16 B. & S. G. Insulated to 1/32.		18 B. & S. G. Insulated to 1/32.	
	Outside diameters. Inches.	Weights, 1000 feet.	Outside diameters. Inches.	Weights, 1000 feet.	Outside diameters. Inches.	Weights, 1000 feet.
2	3/8	102	3/8	92	3/8	82
3	1/2	149	7/16	126	7/16	104
4	9/16	183	1/2	155	7/16	127
5	11/16	226	5/8	198	1/2	151
6	3/4	260	11/16	222	9/16	175
7	13/16	297	3/4	251	5/8	200
10	15/16	401	7/8	335	11/16	256
12	1	465	15/16	393	3/4	296
15	1 1/8	563	1	468	13/16	355
18	1 3/16	651	1 1/16	541	7/8	413
20	1 1/4	714	1 1/8	593	15/16	452
25	1 3/8	868	1 3/16	706	15/16	541
30	1 7/16	1 008	1 1/4	824	1	633
35	1 1/2	1 147	1 5/16	938	1 1/16	723
40	1 9/16	1 268	1 3/8	1 053	1 1/8	813
45	1 5/8	1 431	1 1/2	1 182	1 3/16	903
50	1 3/4	1 577	1 5/8	1 311	1 1/4	994

SUBMARINE CABLES.

Number conductors.	Outside diameters.	Armor wires.		Total weights. Pounds.	
		Number of wires.	Numbers, B. W. G.	1 000 feet.	Mile.
1	⅞	12	8	1 250	6 600
2	1	15	8	1 722	9 092
3	1⅛	14	6	2 363	12 477
4	1 1/16	16	6	2 794	14 752
5	1 1/16	16	6	2 968	15 671
6	1½	16	4	3 822	20 180
7	1½	16	4	3 972	20 972
10	1⅞	18	3	5 404	28 533

The core consists of 7×22 B. & S. tinned copper wires, insulated with rubber to $\frac{5}{32}$ of an inch, laid up with proper jute bedding.

We are prepared to furnish telegraph cables with gutta-percha insulation. This is the best insulation for submarine work, and its reliability and durability more than make up the difference in cost between it and any other insulation.

We are prepared to furnish submarine cables of any description for use in electric lighting and street railway work.

No list of these cables can be made, owing to the varying conditions to be met.

THE COLUMBIA RAIL-BOND.

THE COLUMBIA BOND consists of three parts, two copper thimbles and the connecting copper rod. On each end of this copper rod is a truncated cone-head with a fillet at the base. The inside of the thimble is tapered to fit the head on the bond, while the outside is slightly tapered in the opposite way.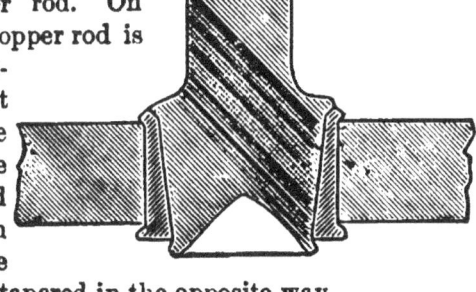

In applying the bond, the cone-shaped heads are placed in the holes in the rail from one side and the thimbles are slipped over them from the other.

A portable hand-press is then applied, and the wedge-shaped head of the bond is forced into the thimble so that it is not possible to see the line separating the thimble and the head in a cross-section of the two.

The end of the head of the bond is expanded by a center-punch, held in position in the press.

When installed, owing to the pressure exerted between the head and the thimble, and also to the fact that they are of the same kind of metal, the two become one, both electrically and mechanically.

The contact of rail and bond is made by a wedge expanding the thimble against the hole in the rail, and, as the bond is wedged both ways, it cannot get loose.

For a 0 000 B. & S. G. or 000 B. & S. G. bond, the holes in the rail should be ⅞-inch, and for a 00 B. & S. G. or a 0 B. & S. G. bond, ⅝-inch.

The total length of a bond is 3½ inches more than the distance from center to center of holes in rails. The total length of a bond should be 8 inches more than that of the splice plate.

www.ingramcontent.com/pod-product-compliance
Lightning Source LLC
Chambersburg PA
CBHW020227090426
42735CB00010B/1615